AF771541

Irreversibilities in Quantum Mechanics

Fundamental Theories of Physics

An International Book Series on The Fundamental Theories of Physics: Their Clarification, Development and Application

Editor:
ALWYN VAN DER MERWE, *University of Denver, U.S.A.*

Editorial Advisory Board:
JAMES T. CUSHING, *University of Notre Dame, U.S.A.*
GIANCARLO GHIRARDI, *University of Trieste, Italy*
LAWRENCE P. HORWITZ, *Tel-Aviv University, Israel*
BRIAN D. JOSEPHSON, *University of Cambridge, U.K.*
CLIVE KILMISTER, *University of London, U.K.*
PEKKA J. LAHTI, *University of Turku, Finland*
ASHER PERES, *Israel Institute of Technology, Israel*
EDUARD PRUGOVECKI, *University of Toronto, Canada*
TONY SUDBURY, *University of York, U.K.*
HANS-JÜRGEN TREDER, *Zentralinstitut für Astrophysik der Akademie der Wissenschaften, Germany*

Volume 113

Irreversibilities in Quantum Mechanics

by

Benjamin Fain
School of Chemistry,
Tel Aviv University,
Tel Aviv, Israel

SPRINGER-SCIENCE+BUSINESS MEDIA, B.V.

A C.I.P. Catalogue record for this book is available from the Library of Congress.

ISBN 978-90-481-5564-4 ISBN 978-0-306-47128-5 (eBook)
DOI 10.1007/978-0-306-47128-5

Printed on acid-free paper

All Rights Reserved
© 2000 Springer Science+Business Media Dordrecht
Originally published by Kluwer Academic Publishers in 2000
Softcover reprint of the hardcover 1st edition 2000
No part of the material protected by this copyright notice may be reproduced or utilized in any form or by any means, electronic or mechanical, including photocopying, recording or by any information storage and retrieval system, without written permission from the copyright owner.

CONTENTS

FOREWORD

The problem of irreversibility is ubiquitous in physics and chemistry. The present book attempts to present a unified theoretical and conceptual framework for the description of various irreversible phenomena in quantum mechanics. In a sense, this book supplements conventional textbooks on quantum mechanics by including the theory of irreversibilities. However, the content and style of this book are more appropriate for a monograph than a textbook.

We have tried to arrange the material so that, as far as possible, the reader need not continually refer elsewhere. The references to the literature make no pretense of completeness. The book is by no means a survey of present theoretical work. We have tried to highlight the basic principles and their results, while the attention has been mainly paid to the problems in which the author himself has been involved. The book as a whole is designed for the reader with knowledge of theoretical physics (especially quantum mechanics) at university level.

This book is based on the courses of lectures given at the Chemistry Department of Tel-Aviv University.

I would like to thank Professors Vitaly Ginzburg and Joshua Jortner for encouraging me to write this book. I am grateful to Shoshana, my wife, for giving me moral support. I would, finally, like to thank Ms. Jacqueline Gorsky for her patience in typing and correcting the manuscript and Ms. Rachel Magen for her highly technical graphical skills in the preparation of the figures.

INTRODUCTION

The following is the structure of the book. Chapter I gives a quantum-mechanical basis and description of the state by the density matrix. Chapter II provides the general theoretical framework for the description of irreversible processes. Chapter III describes irreversible processes due to the interaction with photons, while Chapter IV describes the interaction with phonons. Chapter V provides exact solutions and dissipationless regimes emerging in the system with infinite (in limit) degrees of freedom. Chapter VI, the last one in this book, is devoted to the irreversibilities in quantum measurements.

Chapter I of this book deals with the general scheme of quantum theory. The reader's attention is particularly drawn to the density matrix description of the quantum state. Since our main purpose is the quantum theory of irreversible processes, the use of the density matrix is indispensable. A conventional presentation of quantum mechanics uses the wave function for the description of a quantum system. In this case the density matrix is, in a sense, a derivative of the wave function. As opposed to this we consider the density matrix as a basic, primary characteristic of the quantum system, while the description by the wave function is a specific case of the description with aid of the density matrix. Such an approach is a generalization of the conventional procedure.

The second chapter deals with general theory of irreversible processes. Examples of irreversible processes are a spontaneous emission of an atom, a decay of an electromagnetic field in a lossy resonator, relaxation of a spin system due to the interaction with the crystal lattice, etc. In all these examples the relaxation takes place as a result of the interaction between the dynamic and dissipative system. Atoms, spins, a particular mode of the resonator, etc. present the dynamic system, which has a finite number of degrees of freedom and discrete energy levels. The dissipative system – electromagnetic field in the free space, crystal lattice, etc., has an infinite (in the limit) number of degrees of freedom and a continuous energy spectrum. Interacting dynamic and dissipative systems together form a closed system. The evolution of this closed system is governed by the von Neumann equation. This equation determines the behavior of the whole system. One of the important problems dealt with in Chapter II is the derivation of the equation for the density matrix of the dynamic subsystem, using various approximations.

Chapter III is devoted to the theoretical application of the general formalism developed in Chapter II. The main subject of this application is the interaction of a two-state system (the dynamic system) with the phonon bath (the dissipative system). The quantized vibrations in condensed media or huge molecules may play the role of the pnonon bath. These vibrations have an infinite (in the limit) number of degrees of freedom and a continuum spectrum of energies – the continuum of phonon frequencies. The basic theory of electron and energy transfer, and tunneling in the condensed medium is the main subject of this chapter.

Chapter IV is devoted to the interaction with photons. When eigenfrequencies of photons form the continuum, the interaction with the electromagnetic field vibrations leads to the irreversible processes. These irreversible processes are, to a certain extent, similar to those considered in Chapter III (interaction with phonons).

However, the irreversible processes due to the interaction with photons have their own, peculiar to these processes, features. The relaxation of the system of identical two-level objects (spins, atoms, etc.) may have collective, superradiant character. Therefore, the time development of the superradiance is the main subject of this chapter.

In Chapter V memory effects in the irreversible processes are considered. Most of the results presented in this chapter are exact. The approximate solutions are compared with exact solutions (in a certain range of parameters). Special attention is paid to the dissipationless regimes. The latter emerge in a certain range of parameters of a generally dissipative system, having an infinite number of degrees of freedom and a continuum of energy states.

Various approximations - rotating wave, Markovian and weak coupling are compared with the exact solution for the harmonic oscillator interacting with a harmonic phonon bath.

Chapter VI, the last one, is devoted to the irreversibilities in quantum measurements. The consideration of the irreversibilities in previous chapters shows that only a special type of quantum systems exhibits the irreversibility. Typical finite quantum systems, such as a two-state system, a harmonic oscillator, atoms, and molecules, are reversible. Other kinds of reversible macroscopic systems are superconductors, superfluids and systems considered in chapter V. Nevertheless, chapter VI shows that the irreversibility lies in the very heart of quantum mechanics. Quantum measurement, the process which plays a central role both in the interpretation and essence of quantum mechanics, is irreversible. As opposed to the point of view accepted by many authors, the quantum measurement is considered as a specific type of interaction between the object, measuring device and environment. The role of the observer, the human being, is just to verify the outcomes of the measurement. The irreversible wave function collapse, which is a part of the measurement process, is also described as a special kind of quantum-mechanical interaction. Although the measurement is a physical process, which does not contradict our intuition, there are counterintuitive phenomena connected with the nonlocality of quantum mechanics. These phenomena are also the subject of the considerations in Chapter VI.

In conclusion, this monograph, as mentioned in the foreword, may serve, in a certain sense, as a supplement to conventional textbooks of quantum mechanics, by choosing the following parts of the book: Chapter I, Chapter II, sections 1-4, 6,7, Chapter III, sections 1-5, Chapter IV, sections 1-7, Chapter VI.

CHAPTER 1

QUANTUM-THEORETICAL BASIS. DENSITY MATRIX

The present chapter gives a short account of the basic concepts of quantum theory. Since our main purpose is quantum theory of irreversible processes, the use of the density matrix concept is indispensable. A conventional presentation of quantum theory principles uses the wave function as the basic characteristic of a quantum system. In this case the density matrix is, in a sense, a derivative of the wave function. In this chapter we consider the density matrix as a basic, primary characteristic of the quantum system [1], while the description by the wave function is a specific case of the description with aid of the density matrix. We realize that such an approach is a generalization of the conventional procedure. However, this generalization seems to be quite natural, and only due to historical reasons quantum theory text books use the wave function presentation. In this chapter we follow conventional (Copenhagen) interpretation of quantum mechanics. A different approach will be considered in the last chapter (Chapter VI).

1.1 Basic concepts

The nature of the phenomena occurring at the atomic level is very different from the nature of the phenomena of the macrocosm. For this reason the basic concepts of the classical theory proved to be invalid in describing the microcosm. The concept of the state of a physical system underwent a most radical re-examination. In classical physics it is assumed that the physical quantities (or properties of a system) found from various measurements made on a system are characteristics of the particular state of the system, that they are always present in a given system in a definite form and that this does not depend on the observational methods and equipment. In quantum physics they are at the same time characteristics of the methods and equipment used for the observations. In the microcosm we cannot ignore the effect of the measuring apparatus on the measured object. Therefore the concept of the quantum state takes into account both the object which is in this state and possible experimental devices used to make the measurement. Accordingly, the quantum theoretical description of quantum objects differs essentially from the classical description. Quantum theory, unlike the classical theory, is a statistical theory in principle. The laws of quantum theory do not govern the actual behavior of a particular object, but give the probabilities of the various ways in which the object may behave as a result of an interaction with its surroundings.

The following postulates form the basis of the quantum description of physical phenomena.

1.1.1 Each physical quantity has corresponding to it a linear Hermitian operator or matrix.

For example the radius vector of a particle **r** is associated with the multiplication operator **r**, the momentum of the operator of the particle with the operator $\mathbf{p} = -i\hbar\nabla$, the momentum with the operator $\hbar\mathbf{L} = [\mathbf{r} \times \mathbf{p}] = -i\hbar[\mathbf{r} \times \nabla]$. The operators corresponding to the physical quantities are, generally speaking, not commutative. There are commutation relations between the coordinate and the momentum operators:

$$xp_x - p_x x = i\hbar, \quad yp_y - p_y y = i\hbar, \quad zp_z - p_z z = i\hbar, \tag{1.1.1}$$

and there are also commutation relations between the operators of the components of the angular momentum:

$$L_x L_y - L_y L_x = iL_z, \quad L_y L_z - L_z L_y = iL_x, \quad L_z L_x - L_x L_z = iL_y, \tag{1.1.2}$$

where $\hbar$ is Planck's constant divided by 2π. (Here and further on we do not use special designations for operators, such as $\hat{A}$. The usage of the operators will be clear from the context.) Commutation relations such as (1.1.1) and (1.1.2) are basic characteristics of operators.

1.1.2. Only the eigenvalues of the operator A can be the result of a precise measurement of a physical quantity represented by this operator.
The essential difference from classical theory is the fact that physical quantities may take up a discrete, as well as a continuous, series of values. It is well known, for example, that the energy spectrum of atoms is discrete in nature.

1.1.3. The state of the physical system can be always described by the quantity which is called the density matrix ρ.
The mean value of any physical quantity represented by the operator (matrix) A is given by the formula

$$\langle A \rangle = \mathrm{Tr}(\rho A) \equiv \sum_{n,m} \rho_{nm} A_{mn}, \tag{1.1.3}$$

where A_{mn}, ρ_{nm} are matrix elements of operators A and ρ respectively. These matrix elements are taken over a full set of eigenfunctions of some Hermitian linear operator, and Tr means the trace of the matrix.

1.1.4 The time evolution of quantum systems is described by the von Neumann equation

$$i\hbar\frac{\partial\rho}{\partial t} = [H, \rho], \tag{1.1.4}$$

where H is a Hamiltonian of the system.

These postulates have sense provided the correspondence is established between physical quantity and its operator. The first postulate has to be understood in the sense that such a correspondence is realized.

Now let us consider some implications from the above postulates and particularly a statistical description of quantum systems by the density matrix.

1.2 Various representations of physical quantities.

It follows from the above postulates that both the states of the physical system and physical quantities themselves are described by matrices. These matrices ρ and A are second-rank tensors in a complex space of infinite number of dimensions – in so-called Hilbert space. However, in order to realize an analogy with tensors it is necessary to establish the law of matrix transformation.

Let ψ_n be eigenfunctions of some linear Hermitian operator F

$$F\psi_n = f_n\psi_n \quad , \tag{1.2.1}$$

where f_n is the eigenvalue of the operator F. Such a system of eigenfunctions is complete and can be chosen orthonormal. Matrix elements of operator A in this basis can be presented as

$$A_{mn} = <m|A|n> = \int \psi_m^*(q) A \psi_n(q) dq \quad , \tag{1.2.2}$$

where q – is a set of arguments of function ψ_n. Matrix elements (1.2.2) give the F-presentation of operator A. The same physical quantity in another G- representation is presented by matrix

$$A'_{nm} = \int \psi_m'^*(q) A \psi_n'(q) dq \quad , \tag{1.2.3}$$

where ψ_n' are eigenfunctions of another operator G. They present another basis of eigenfunctions. The law of transformation of matrices A from one basis to another is determined by the transformation of functions $\psi_n \to \psi_n'$, i.e. by transformation of vectors ψ_n in Hilbert space. This is a well-known unitary transformation U. Matrix elements of operator U are determined by the expansion of function ψ_n' over a complete set of functions ψ_m

$$\psi_n' = \sum_m U_{mn}\psi_m = U\psi_n \quad . \tag{1.2.4}$$

Now we can find connection between matrix elements in G-representation - A'_{mn} and those in F-representation - A_{mn}

$$A'_{mn} = \int \psi_m'^* A \psi_n' dq = \int (U\psi_m)^* A (U\psi_n) dq$$

$$= \sum_{k\ell} \int U^*_{km} \psi^*_k A U_{\ell n} \psi_\ell dq = \sum_{k\ell} U^*_{km} A_{k\ell} U_{\ell n}$$
$$= \sum_{k,\ell} U^+_{mk} A_{k\ell} U_{\ell n} = (U^+ AU)_{mn} \quad , \tag{1.2.5}$$

where the Hermitian conjugated operator U^+ is defined by the relation

$$(U^+)_{mk} = U^*_{km} \equiv \tilde{U}^*_{mk} \quad , \tag{1.2.6}$$

and the rule of the multiplication of matrices is

$$(AB)_{nm} = \sum_k A_{nk} B_{km} \quad . \tag{1.2.7}$$

From the fact that eigenfunctions of the Hermitian operators can be chosen orthogonal and normalized to unity, the unitary property of the operator U follows

$$\int \psi'^*_n \psi'_m dq = \int \sum_{k\ell} U^*_{kn} \psi_k U_{\ell m} \psi_\ell dq = \sum_{k,\ell} U^+_{nk} U_{\ell m} \delta_{k\ell}$$
$$= \sum_k U^+_{nk} U_{km} = (U^+ U)_{nm} = \delta_{nm} \quad . \tag{1.2.8}$$

Thus, the unitary operator has the property

$$U^+ U = 1; \qquad U^+ = U^{-1} \quad , \tag{1.2.9}$$

where 1 is a unit operator with matrix elements δ_{nm}. The transformation property of the Hermitian operator A, representing some physical quantity, can be found from Eq. (1.2.5)

$$A' = U^+ AU = U^{-1} AU \quad . \tag{1.2.10}$$

Transformation properties of the density matrix can be determined from the requirement that mean values of the physical quantities are scalars – invariants of the unitary transformation. This requirement is quite natural, mean values of physical quantities should not depend on the choice of the basis of the eigenfunctions. Therefore,

$$< A' > = < A > = \mathrm{Tr}\rho' A' = \mathrm{Tr}\rho' U^+ AU = \mathrm{Tr}\rho A \quad . \tag{1.2.11}$$

Using the property of a trace of the product of two matrices

$$\mathrm{Tr}AB = \mathrm{Tr}BA \quad , \tag{1.2.12}$$

one can get from Eq. (1.2.11)

$$< A >= \mathrm{Tr} U\rho' U^{+} A = \mathrm{Tr}\rho A \quad , \tag{1.2.13}$$

Therefore, the density matrix has the transformation property of a second rank tensor

$$\rho = U\rho' U^{+}; \qquad \rho' = U^{+}\rho U = U^{-1}\rho U \quad . \tag{1.2.14}$$

Thus the same physical state is represented by all density matrices ρ connected by transformation (1.2.14). Notice that the fact that the mean value (1.2.11) is real, implies the Hermiticity of the density matrix

$$\rho_{mn} = \rho^{*}_{nm} \quad . \tag{1.2.15}$$

1.3 Quantum state and statistical ensembles

Let us enter in more detailed description of quantum state and its characteristic – the density matrix. First of all we establish statistical ensembles determining mean values of physical quantities. In the conventional interpretation of quantum mechanics the quantum state relates to one object. To verify that certain density matrix describes the state of the object one has to perform a large number of measurements. As it has been mentioned above we cannot ignore the effect of the measuring device on the measured object. A measurement or experiment, generally speaking, changes the state of the object. It is therefore necessary (in order to remain in the same quantum state) to return the object after each measurement to the original quantum state or to deal with a set of objects in one and the same quantum state (quantum state is described by the density matrix ρ). In the latter case the measurement is made once on each object.

Having the set of objects which are in the same quantum state, determined by the density matrix ρ, we can perform different kinds of measurements. Thus we can measure an electron coordinate, its momentum, spin, etc. Each kind of measurement produces its statistical ensemble. In the ensemble produced in this way we can introduce the probability distribution for obtaining a particular result of a measurement. Therefore to define a statistical ensemble in quantum theory we must first have the state of the object (its density matrix ρ), and secondly choose the type of measurement to be made on the object. Thus, a statistical ensemble in quantum theory is a set of identical measurements made on an object in a given quantum state.

The question arises how to find probability distribution in various ensembles produced by the measurements of various quantities in the state described by the density matrix ρ. To answer this question we write the mean value of physical quantity A with the aid of formula (1.1.3) using A-representation. In this representation

$$A\psi_n = a_n \psi_n \quad , \tag{1.3.1}$$

and

$$A_{mn} = a_n \delta_{mn} \quad , \tag{1.3.2}$$

where a_n are the eigenvalues of the operator A. The mean value of quantity A takes the form

$$< A >= \sum_n \rho_{nn} a_n \quad . \tag{1.3.3}$$

In formula (1.3.3), it is assumed that density matrix ρ is normalized to the unity

$$\mathrm{Tr}\rho = \sum_n \rho_{nn} = 1 \quad , \tag{1.3.4}$$

(If the density matrix is not normalized to unity it can always be normalized multiplying it by the factor $(\sum \rho_{nn})^{-1}$).

From Eqs. (1.3.3) and (1.3.4) and second postulate it follows that ρ_{nn} gives the probability to get a_n, while measuring quantity A. To get the probability distribution of another quantity, say F, we have to pass to F-representation and diagonal elements of ρ in this representation will give distribution of probabilities to get eigenvalues f_n.

Thus probability distributions of various kinds of measurements, performed over an object in quantum state ρ, are connected by the unitary transformation U, (1.2.14),

$$\rho'_{nn} = \sum_{km} (U^{-1})_{nk} \rho_{km} U_{mn} ; \qquad \rho_{mm} = \sum_{l,k} U_{ml} \rho'_{lk} (U^{-1})_{km} \quad . \tag{1.3.5}$$

It is worthwhile to emphasize that all representations of density matrix (i.e. various reference systems in the Hilbert space) are equivalent. Statistical ensemble is determined by the kind of measurement performed over a system in quantum state ρ.

1.4 The wave function

Up till now we did not use the notion of the wave function. As is known, originally quantum mechanics was based on this notion and the density matrix was introduced later. Description of quantum state in text-books is based on the wave function. It means that all statistical characteristics of the system can be derived from the wave function ψ(q). It is assumed that the average value of any physical quantity can be obtained by the formula

$$< A >= \int \psi^*(q) A \psi(q) dq \quad . \tag{1.4.1}$$

This formula is different from Eq. (1.1.3). We will see that while formula (1.1.3) and the density matrix describe all possible quantum states, formula (1.4.1) and the wave function describe very specific kinds of quantum state which are called pure states. Later on we will show that pure states are the most ordered and most coherent quantum states.

First of all we want to show that there are quantum states which cannot be described by the wave function. It means that various statistics of such states cannot be derived from formula (1.4.1). The fact that a state cannot always be described by the wave function can be understood if we examine the subsystem A of a certain system A+B. Let the system A+B be described by the wave function $\psi = \psi(x_A, x_B)$, where x_A and x_B are the coordinates of the subsystems A and B respectively. This function, generally speaking, does not break down into the product of the functions ψ_A and ψ_B even if the systems A and B do not interact. In the case when function $\psi(x_A, x_B)$ is factorized into a product of two functions

$$\psi(x_A, x_B) = \psi_A(x_A)\psi_B(x_B) \quad , \tag{1.4.2}$$

each of the subsystems A and B can be described by the wave function. In this case the mean value of some quantity A of the subsystem A has the form

$$\begin{aligned} <A> &= \int \psi^*(x_A, x_B) A \psi(x_A, x_B) dx_A dx_B \\ &= \int \psi_A^*(x_A) A \psi_A(x_A) dx_A \int \psi_B^*(x_B)\psi_B(x_B) dx_B \\ &= \int \psi_A^*(x_A) A \psi_A(x_A) dx_A \quad . \end{aligned} \tag{1.4.3}$$

Thus we have shown that subsystem A can be described by the wave function ψ_A. (It has been assumed here that wave functions are normalized to unity). In the same way we can show that subsystem B is described by the wave function $\psi_B(x_B)$.

In the general case we cannot describe system A (or B) by the wave function, even if there is no interaction between systems A and B. To exemplify this situation (see Fig. 1), let us consider two subsystems A and B to be non-interacting at $t = -\infty$, and the wave function of the combined system A+B is factorized at $t = -\infty$ according to Eq. (1.4.2). At time $t = 0$ the interaction (collision) takes place between subsystem A and B which lasts time $\sim t_c$. Let ψ_{An} and ψ_{Bl} be eigenfunctions of energies of sybsystems A and B, respectively, ignoring interactions between them. Then, in the general case, the wave function of combined system A+B can be written in the form

$$\psi(x_A, x_B, t) = \sum_{n,\ell} c_{n\ell}(t) \psi_{An}(x_A) \psi_{B\ell}(x_B) e^{-\frac{i}{\hbar}(E_n^A + E_\ell^B)t} \quad , \tag{1.4.4}$$

where E_n^A and E_ℓ^B are eigenenergies of subsystems A and B respectively, while $c_{n\ell}(t)$ are time-dependent coefficients. The wave function of the combined system A+B satisfies the Schrődinger equation

$$i\hbar \frac{\partial \psi}{\partial t} = (H_A + H_B + V_{AB})\psi \quad . \tag{1.4.5}$$

At time t much larger than the collision time t_c

$$t \gg t_c \ , \tag{1.4.6}$$

we may ignore the interaction V_{AB} between subsystems A and B. Two subsystems are conservative, described by Hamiltonians H_A and H_B respectively. The whole system A+B is described by the wave function

$$\psi = \sum_{n,\ell} c_{n\ell} \psi_{An} \psi_{B\ell} e^{-\frac{i}{\hbar}(E_n^A + E_\ell^B)t} \ , \tag{1.4.7}$$

where c_{nl} are time-independent coefficients. Now neither of non-interacting, conservative systems A and B can be described by the wave function. It means that statistical characteristics of these systems cannot be described according to formula (1.4.1). The question what is the wave function of system A (or B) does not make sense.

Thus we come to the conclusion that in the general case the quantum system (both isolated and interacting with other systems) cannot be described by the wave function. Nevertheless both mixed states (not described by the wave function) and pure states can be described, in the unified way, by the density matrix. Let us first find the density matrix of the pure state described by the wave functions

$$\psi(q) = \sum_n b_n \psi_n(q) \ , \tag{1.4.8}$$

where ψ_n are eigenfunctions of some linear Hermitian operator. The mean value of some physical quantity may be written in the form

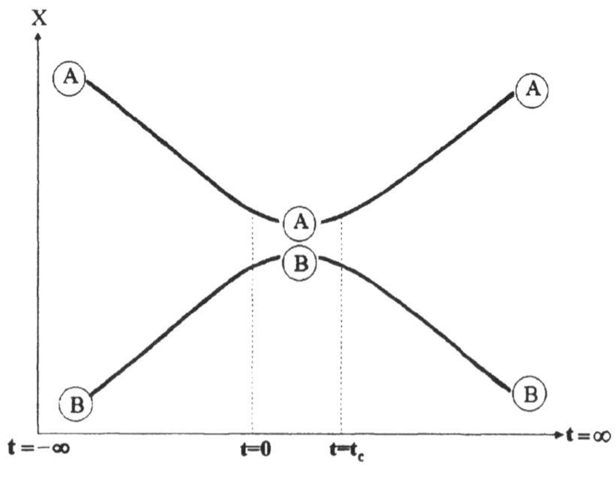

Figure 1. Two subsystems A and B are statistically independent and noninteracting at $t = -\infty$ (x is the distance between the systems). At time $t = 0$ they collide, t_c is the collision time. At $t \to \infty$ A and B do not interact, but they are statistically dependent.

$$<A>=\int\psi^*A\psi dq=\sum_{nn'}A_{nn'}b_{n'}b_n^* \quad . \tag{1.4.9}$$

Denoting

$$\rho_{n'n}=b_n^*b_{n'} \quad , \tag{1.4.10}$$

we can rewrite the expression for the mean value of A (in a more symmetrical form)

$$<A>=\sum_{n'n}\rho_{n'n}A_{nn'}=\mathrm{Tr}(\rho A) \quad . \tag{1.4.11}$$

According to Eqs. (1.1.3) and (1.4.11), the quantity $\rho_{n'n}$, Eq. (1.4.10), is the density matrix of the pure state described by the wave function (1.4.8).

Now let us consider the subsystem A of the combined system described by the wave function (1.4.4). A certain operator A (describing physical quantity A) of the subsystem A has matrix elements

$$<n,k\,|\,A\,|\,n',k'>=A_{nn'}\delta_{kk'} \quad , \tag{1.4.12}$$

where n,n' are indices of subsystem A, while indices k,k' describe subsystem B. Since the whole system is in a pure state described by the wave function (1.4.4), its density matrix according to Eq. (1.4.10) is

$$\rho_{nk;n'k'}=c^*_{n'k'}c_{nk}e^{-\frac{i}{\hbar}(E_n^A-E_{n'}^A)t}e^{-\frac{i}{\hbar}(E_k^B-E_{k'}^B)t} \quad , \tag{1.4.13}$$

The mean value of quantity A is [according to Eqs. (1.4.11) and (1.4.12)]

$$<A>=\sum_{n,k,n',k'}\rho_{nk;n'k'}A_{n'n}\delta_{kk'}=\sum_{n,n',k}\rho_{nk;n'k}A_{n'n}=\sum_{nn'}\sigma_{nn'}A_{n'n} \quad , \tag{1.4.14}$$

where

$$\sigma_{nn'}=\sum_k\rho_{nk;n'k} \quad , \tag{1.4.15}$$

or

$$\sigma=\mathrm{Tr}_B\rho \quad . \tag{1.4.16}$$

The matrix σ is by definition the density of the subsystem A. As it was mentioned above, the subsystem A does not have the wave function, while its statistical properties can be described by density matrix σ. As it is clear from the derivation of

Eqs. (1.4.15) and (1.4.16), they are valid also for the case when the combined system A+B is not described by the wave function and is described by some density matrix of a general kind.

When systems A and B do not interact, coefficients c_{nk} are time-independent. In this case the time dependence of the subsystem A has the form

$$\sigma_{nn'}(t) = \sigma_{nn'}(0)e^{-\frac{i}{\hbar}(E_n^A - E_{n'}^A)t} , \tag{1.4.17}$$

and $\sigma_{nn'}(t)$ satisfies the von Neumann equation (1.1.4).

1.5 Entropy of quantum ensembles

As it has been mentioned above the wave function describes a very specific kind of quantum states, namely, the most ordered (non-chaotic) states – so called pure states. To talk about statistical properties of the state, we should have some measure of these properties. Such a measure is provided to us by the information theory and is called the entropy of ensemble [2]. (Of course, this entropy is a generalization of the entropy used in statistical mechanics.) The entropy of the ensemble is a measure of the statistical scatter or chaotical nature of the probability distribution in the ensemble.

By the definition, the entropy satisfies the following conditions. It is a functional of the probability distribution which has its maximal values in the most chaotic ensemble in which all possible values of measured quantity has equal probabilities

$$P_1 = P_2 = \ldots = P_n = \ldots = P_N = \frac{1}{N} . \tag{1.5.1}$$

Here n may be the quantum number of energy (E_n are eigenenergies), spin, angular momentum, etc. The entropy has its minimal value (zero) when the measured quantity has a definite value, i.e. the probability of certain quantum number n_o is equal to unity

$$P_{n_o} = 1 . \tag{1.5.2}$$

And, lastly, the entropy must be additive: the entropy of a system consisting of two statistically independent subsystems is equal to the sum of the entropies of each subsystem. All these conditions (except for the inclusion of a constant factor) are satisfied by the quantity

$$E = -\sum_i P_i \ln P_i , \tag{1.5.3}$$

where P_i is the probability of the i-th value (quantum number) of the measured quantity. The entropy of the most chaotic ensemble, (1.5.1), takes the value

$$E = -\sum_i \frac{1}{N} \ln \frac{1}{N} = \ln N \ , \qquad (1.5.4)$$

while the entropy of the most ordered ensemble equals

$$E = -P_{n_o} \ln P_{n_o} = 0 \ , \qquad (P_{n_o} = 1) \ . \qquad (1.5.5)$$

Now let us return to the statistical ensembles corresponding to some quantum state with the density matrix ρ. As we know, there are many ensembles corresponding to this state, but we always can find the less chaotic ensemble – i.e., with minimal entropy. This may be done with the aid of the mathematical theorem that states: in the representation in which ρ is diagonal, the sum

$$\sum_n \rho_{nn} \ln \rho_{nn} \ , \qquad (\rho_{nm} = 0, \ \ n \neq m) \ ,$$

is larger than in any other representation with $\rho'_{nm} \neq 0$ $(n \neq m)$

$$\sum_n \rho_{nn} \ln \rho_{nn} > \sum_n \rho'_{nn} \ln \rho'_{nn} \ .$$

Thus the quantity

$$E = -\sum_n \rho_{nn} \ln \rho_{nn} \equiv -\mathrm{Tr}(\rho \ln \rho) \ , \qquad (1.5.6)$$

may serve as a measure of the statistical properties of the state ρ: it is the entropy of the most ordered ensemble (with minimal entropy) among the ensembles of state ρ.

1.6 Pure and mixed states. Proper mixtures.

Thus we have introduced a statistic characteristic of the quantum state ρ. This is the entropy of the less chaotic ensemble corresponding to state ρ. Now we can compare statistical properties of various quantum states. It is expedient to distinguish the special class of states having the entropy equal to zero. These are the most ordered states. According to Eq. (1.5.5), it is easy to see that for these states, density matrix ρ_{nm}, in the representation in which it is diagonal, is equal

$$\rho_{nm} = \delta_{n_o n} \delta_{n_o m} \ . \qquad (1.6.1)$$

In the arbitrary representation

$$\rho^2 = \rho \ , \qquad (1.6.2)$$

which, together with the equation

$$\mathrm{Tr}(\rho \ln \rho) = 0 \quad , \tag{1.6.3}$$

characterize the class of the pure states.

From Eq. (1.6.1) follows an important property of the density matrix of pure states. In the arbitrary representation this density matrix may be factorized

$$\rho_{nm} = c_n c_m^* \quad . \tag{1.6.4}$$

To prove this property we employ the transformation rule (1.2.14) and Eq. (1.6.1)

$$\rho'_{nm} = (U^{-1})_{nn_0} U_{n_0 m} \quad . \tag{1.6.5}$$

Designating

$$(U^{-1})_{nn_0} = c_n \quad , \tag{1.6.6}$$

we obtain Eq. (1.6.4).

Thus, the pure state may be characterized by the set of numbers c_n. This set of numbers is usually called the wave vector. It transforms as a vector in the Hilbert space, with the aid of the unitary transformation (1.2.4). This vector characterizes the pure state, together with the density matrix (1.6.4). The mean value of the arbitrary operator A may be determined as

$$< A >= \sum_{mn} \rho_{nm} A_{mn} = \sum_{mn} c_m^* A_{mn} c_n = < c \mid A \mid c > \quad . \tag{1.6.7}$$

In particular when indices n take on a continuous set of values (e.g., coordinates of a particle) we call c_m a wave function, designate it as $\psi(q)$ and instead of the summation in Eq. (1.6.7) we perform the integration

$$< A >= \int \psi^*(q) A \psi(q) dq \, , \quad \int \mid \psi(q) \mid^2 dq = 1 \quad . \tag{1.6.8}$$

The density matrix in this case is equal to

$$\rho(q, q') = \psi(q) \psi^*(q') \quad , \tag{1.6.9}$$

and according to the general rule the probability distribution takes the form

$$P(q) = \rho(q, q) = \mid \psi(q) \mid^2 \quad . \tag{1.6.10}$$

The pure state, being the most ordered state, is also, in a sense, the most coherent state. To show this we transform density matrix ρ_{mn} into a q-representation. Unitary transformation from the n-representation to the q-representation can be written in the form similar to Eq. (1.2.4)

$$\psi(q) = \sum_n U_{nq} c_n = \sum_n c_n \psi_n(q) \ , \tag{1.6.11}$$

$\psi_n(q)$ are orthogonal and normalized eigenfunctions of some linear Hermitian operator; they are, at the same time, matrix elements of unitary transformation from n- to q-representation. Therefore, the density matrix in the q-representation obtains the form

$$\rho(q,q') = \sum_{nm} \rho_{nm} \psi_n(q) \psi_m^*(q') \ . \tag{1.6.12}$$

Thus, the probability distribution of coordinates q takes the form

$$\rho(q,q) = \sum_n \rho_{nn} |\psi_n(q)|^2 + \sum_{n \neq m} \rho_{nm} \psi_n(q) \psi_m^*(q) \ . \tag{1.6.13}$$

As is known the coherence is usually connected with the possibility of obtaining the interference picture. The interference picture is connected with superposition of two fields (particularly, optical fields). In the specific case when $\psi_n(q)$ are electron de Broglie waves, Eq. (1.6.13) describes the electron diffraction.

The coherence is characterized by off-diagonal elements ρ_{nm}. A quantity

$$g_{nm} = \frac{\rho_{nm}}{(\rho_{nn}\rho_{mm})^{1/2}} \ , \tag{1.6.14}$$

is a natural measure of the interference contrast and, therefore, of the coherence. It can be shown that quantities ρ_{nm} satisfy inequality

$$|\rho_{nm}|^2 \leq \rho_{nn}\rho_{mm} \ . \tag{1.6.15}$$

The maximal degree of the coherence is characterized by

$$|g_{nm}| = 1 \ , \tag{1.6.16}$$

and is achieved in the case of pure state when $\rho_{nm} = c_n c_m^*$, Eq. (1.6.4). All other, so called mixed states are characterized by $|g_{nm}| < 1$, and, particularly when g_{nm} equals zero – the interference picture is entirely absent. Thus pure states have maximal coherence – they could be called coherent states.

Another example is a molecule (or atom) interacting with an electromagnetic field. We assume that only two low lying levels E_a and E_b are relevant to the interaction with the electromagnetic field. In the dipole approximation, a dipole moment of the molecule determines the interaction with the field. The mean value of the dipole moment has the form

$$<d>=\rho_{aa}d_{aa}+\rho_{bb}d_{bb}+\rho_{ab}d_{ba}e^{i\omega_0 t}+\rho_{ba}d_{ab}e^{-i\omega_0 t} \quad , \tag{1.6.17}$$

where d_{mn} are matrix elements of the dipole moment (m,n = a,b),

$$\omega_0 = \frac{E_b - E_a}{\hbar} \approx \omega \quad , \tag{1.6.18}$$

and ω is the frequency of the electromagnetic field.

Only time-dependent terms in Eq. (1.6.17) determine interaction with the electromagnetic field. Matrix elements $\rho_{ab} = \rho^*_{ba}$ determine coherent time- dependence of the mean dipole moment, Eq. (1.6.2). Thus the measure of the coherence in this case is

$$g_{ab} = \frac{\rho_{ab}}{\sqrt{\rho_{aa}}\sqrt{\rho_{bb}}} \quad . \tag{1.6.19}$$

Again in the pure case $|g_{ab}| = 1$, while in the most chaotic example

$$g_{ab} = 0$$

We continue discussing pure and mixed states. According to Eq. (1.6.1) the density matrix of the pure state (in the diagonal representation) has the form

$$\hat{\sigma}_n = \begin{pmatrix} 0 & & & & \\ & 0 & & 0 & \\ & & 1 & & \\ & 0 & & 0 & \\ & & & & 0 \end{pmatrix} \tag{1.6.20}$$

It is obvious that the density matrix of the most general mixed state is equal to the linear superposition of the possible matrices of $\hat{\sigma}_n$ of pure states

$$\rho = \sum p_n \hat{\sigma}_n \quad , \tag{1.6.21}$$

where

$$\sum_n p_n = 1; \qquad p_n \geq 0 \quad . \tag{1.6.22}$$

Condition (1.6.22) ensures the normalization of ρ to 1. Carrying out the unitary transformation of all the matrices $\hat{\sigma}_n$ in (1.6.21) we come to the most general mixed state density matrix.

If the object is in a pure state with a given wave function, this wave function describes the behavior of this individual object. It characterizes potential properties of the object which are realized when the measurement is made on the object when it is in definite external conditions. If, however, the object is in a mixed state and has no wave function, it is described by the density matrix, which is also a characteristic of this individual object. Again the density matrix describes potential properties of the individual object.

Let us take a simple example which illustrates the use of the density matrix. Let a system consisting of two spins ½ be in a pure state with a total spin of zero (i.e. all the spin components and its absolute value are equal to zero). As is well known (see e.g. Landau and Lifshitz [3]) the wave function of a state of this kind is of the form

$$\psi_{0,0} = \frac{1}{\sqrt{2}}\left\{\psi_{\frac{1}{2}}(1)\psi_{-\frac{1}{2}}(2) - \psi_{-\frac{1}{2}}(1)\psi_{\frac{1}{2}}(2)\right\} , \qquad (1.6.23)$$

where $\psi_{\pm\frac{1}{2}}$ are the eigenfunctions of the operator of the projection of an individual spin onto the z-axis.

We shall now examine the state of the first spin only, without taking any notice of the state of the second. The first spin is supposed to be a separate system which has no wave function and is in a mixed state. The density matrix of the state can be obtained from the density matrix of the whole system by using (1.4.15) and (1.4.16). The matrix elements of the density matrix of the whole system are of the form

$$\sigma_{\frac{1}{2},-\frac{1}{2};\frac{1}{2},-\frac{1}{2}} = \frac{1}{2}; \quad \sigma_{-\frac{1}{2},\frac{1}{2};-\frac{1}{2},\frac{1}{2}} = \frac{1}{2} , \qquad (1.6.24)$$

$$\sigma_{\frac{1}{2},-\frac{1}{2};-\frac{1}{2},\frac{1}{2}} = -\frac{1}{2}; \sigma_{-\frac{1}{2},\frac{1}{2};\frac{1}{2},-\frac{1}{2}} = -\frac{1}{2} . \qquad (1.6.25)$$

where the first index denotes the state of the first spin and the second one the state of the second spin. Retaining only diagonals for the second spin we can use Eq. (1.4.15) to find the density matrix of the first spin in the form

$$\hat{\rho} = \frac{1}{2}\begin{pmatrix} 1 & 0 \\ 0 & 1 \end{pmatrix} . \qquad (1.6.26)$$

It follows from this, that measurement of the projection of the spin onto the z-axis with a probability of ½ leads to a value of ½, and with a probability of ½ to a value of -½. Before measurement, however, there was no definite value for the projection of the first spin onto the z-axis. This can be seen from Eq. (1.6.23): the spin was in a superposition of the states ½ and –½.

It should be stressed that since Eq. (1.6.26) describes the state of the individual spin, it cannot be interpreted by saying that the projection of the spin onto the z-axis has a definite value, but <u>we do not know it</u>. Let us assume that in the state (1.6.26) the z-component of the spin has in fact a definite, but unknown, value. We now change to another representation in which the x-component of the spin is diagonal. In this representation the density matrix will have the same form (1.6.26). (The matrix (1.6.26) is proportional to the unit matrix and the latter, as is well known, is of exactly the same form in all representations.) Then the density matrix elements are the probabilities for the x-component of the spin. Just as for the case of the z-component, we should have assumed that the x-component of the spin has a definite value but we do not know it. A similar argument can be adduced for the y-component. Therefore the assumption that (1.6.26) in the general case describes a definite, but unknown, value of the component of the individual spin leads to the result that is absurd from the point of view of quantum mechanics that all three components of the spin have definite values. This contradicts the uncertainty relation for the components of the spin.

It has been stressed that matrix (1.6.26) contains the information about a <u>separate</u> spin. However, the question arises whether the first spin can be considered separately from the second spin. More in detail we will consider this problem in Chapter VI. Here we only mention that matrix (1.6.26) gives <u>only partial</u> information about the first spin. This information is derived from the matrix elements (1.6.24). However, matrix elements (1.6.25) also contain information about the first spin as a part of a two-spin system. These matrix elements describe statistical correlations between the first and second system.

Very often the notion of a mixed state is used in a different sense. It is considered as a <u>mixture</u> of pure states. In this case the density matrix describes an ensemble of objects (and not an individual object), each of which is in a definite state. Such density matrix describes the statistical features connected with our <u>lack of knowledge</u> of certain properties of the objects, although these properties are peculiar to the object in a definite form. In fact let us examine a set of objects each of which has a probability p_n of being in one of the pure states with a density matrix $\hat{\sigma}_n$ from (1.6.20). Then all the mean values in this ensemble are defined by the density matrix (1.6.21) in accordance with (1.1.3)

$$< A >= \mathrm{Tr}(A\rho) = \sum_n p_n \mathrm{Tr}(A\hat{\sigma}_n) = \sum_n p_n < A_n > . \tag{1.6.27}$$

where $<A_n>$ is the mean value of quantity A in pure state $\hat{\sigma}_n$. Thus the mean value in the ensemble of pure states (mixture of states) is equal to the sum of the mean values in the state $\hat{\sigma}_n$ multiplied by the probability of this state in the ensemble. Sometimes such a mixture of states is called the <u>proper</u> mixture (d'Espagnat [4]).

The proper mixture has to be distinguished from the mixed state describing the individual quantum object. Let us consider the proper mixture of states of spin ½. Let the probability of the pure state $s_z = ½$ be ½ and the probability of state $s_z = -½$ be ½. Then the density matrix of such an ensemble takes the form

$$\rho=\frac{1}{2}\left(\hat{\sigma}_{\frac{1}{2}}+\hat{\sigma}_{-\frac{1}{2}}\right)=\frac{1}{2}\begin{pmatrix}1 & 0\\ 0 & 0\end{pmatrix}+\frac{1}{2}\begin{pmatrix}0 & 0\\ 0 & 1\end{pmatrix}=\frac{1}{2}\begin{pmatrix}1 & 0\\ 0 & 1\end{pmatrix}. \tag{1.6.28}$$

In another representation with the basis $\psi'_{\pm\frac{1}{2}}$ (with $s_x=\pm\frac{1}{2}$) the density matrix takes the form

$$\rho'=\frac{1}{2}\left(\hat{\sigma}'_{\frac{1}{2}}+\hat{\sigma}'_{-\frac{1}{2}}\right)=\frac{1}{2}\frac{1}{2}\begin{pmatrix}1 & 1\\ 1 & 1\end{pmatrix}+\frac{1}{2}\frac{1}{2}\begin{pmatrix}1 & -1\\ -1 & 1\end{pmatrix}=\frac{1}{2}\begin{pmatrix}1 & 0\\ 0 & 1\end{pmatrix}. \tag{1.6.29}$$

While the density matrix (1.6.28) gives the probabilities ½ for states $s_z = \pm\frac{1}{2}$, the density matrix (1.6.29) provides probabilities ½ for states $s_x = \pm\frac{1}{2}$. It has to be noticed that in this proper mixture, s_z-representation of the density matrix gives the probabilities of states $s_z = \frac{1}{2}$, these states are peculiar to the objects in a definite form. On the other hand density matrix ρ', (1.6.29), provides probabilities ½ to find object $s_x = \pm\frac{1}{2}$ as a result of the measurement, while before the measurement spins did not have definite s_x values. Thus in the proper mixture, not all the representations are equivalent. Only s_z-representations give the probabilities to find spins in a definite form (which they have before the measurement). In all other representations, s_x, s_y (and other axes) probabilities ½ correspond to values of spin components resulting from the measurements of these components.

1.7 Transition probability per unit time

The notion of the probability per unit time plays a very important role in the theory of irreversible processes. Therefore it is expedient to dwell upon this notion more in detail. A conventional description of this concept is sometimes misleading.

First we develop a non-stationary perturbation theory using the density matrix formalism. Let us consider a quantum mechanical system described by the Hamiltonian

$$H=\mathrm{H}_0+\mathrm{V}\ , \tag{1.7.1}$$

where V is a part of the whole Hamiltonian which may be considered as a small perturbation. The meaning of the word "small" would be clear from further derivation. According to the von Neumann equations (1.1.4) we have

$$i\hbar\frac{\partial\rho}{\partial t}=[H_0+V,\rho]\ . \tag{1.7.2}$$

If V is a small perturbation it is expedient to use the interaction representation. This representation can be achieved by unitary transformation

$$A_{int}=U_0^{-1}AU_0\,;\qquad V_{int}=U_0^{-1}VU_0\,;\qquad U_0=\exp\left(-\frac{i}{\hbar}H_0t\right)\ . \tag{1.7.3}$$

The density matrix in this representation takes the form

$$\rho_{int}=U_0^{-1}\rho U_0\ , \tag{1.7.4}$$

It is easy to check that the density matrix in the interaction representation takes the form

$$i\hbar\frac{\partial\rho}{\partial t}=[V,\rho]\ . \tag{1.7.5}$$

Here and further on we omit index "int".

We start from equation (1.7.5) and consider V to be a small quantity. We expand the density matrix into a series

$$\rho=\sum_k\rho^{(k)}\ , \tag{1.7.6}$$

where $\rho^{(k)}$ is a quantity of the k-th order of smallness with respect to V. Substituting (1.7.6) into (1.7.5) we obtain the system of recurrence relations

$$i\hbar\frac{\partial\rho^{(k)}}{\partial t}=\left[V,\rho^{(k-1)}\right]\ . \tag{1.7.7}$$

As the zero approximation we shall take the value of the density matrix at t = 0. This means that the corrections of the first and subsequent approximations of $\rho^{(k)}$ should be small in comparison with $\rho(0)$. We write the density matrix explicitly with an accuracy up to terms of the second order of smallness

$$\rho(t)=\rho^{(0)}+\rho^{(1)}+\rho^{(2)}+\ldots=\rho(0)-\frac{i}{\hbar}\int_0^t[V(t_1),\rho(0)]dt_1$$

$$-\frac{1}{\hbar^2}\int_0^t dt_1\int_0^{t_1}dt_2\left[V(t_1),[V(t_2),\rho(0)]\right]+\ldots\ . \tag{1.7.8}$$

When V does not depend on time (in the Schrödinger picture), time-dependence V(t) in Eqs. (1.7.5), (1.7.7), and (1.7.8) is determined by second relation (1.7.3). In general, the perturbation V may also have an explicit time-dependence. This explicit time-dependence corresponds to time-dependent forces acting on the unperturbed system.

Now let us examine the special case when initially the system is in the stationary state n_0 of the unperturbed Hamiltonian H_0

$$\rho_{mn}(0) = \delta_{mn_o}\delta_{nn_o} \quad . \tag{1.7.9}$$

Substituting Eq. (1.7.9) into the right-hand side of (1.7.8) (and considering the case when V is explicitly time-independent) we find

$$\rho_{nn}(t) = \frac{2|V_{nn_o}|^2}{(E_{n_o} - E_n)^2}\left[1 - \cos(E_{n_o} - E_n)\frac{t}{\hbar}\right], \quad (n_o \neq n) \quad , \tag{1.7.10}$$

where E_n, E_{n_o} are the energy levels of the unperturbed system. The expression (1.7.10) for $\rho_{nn}(t)$ is the probability of finding (when measuring) the system in the state n if it was in the state n_o initially. In other words $\rho_{nn}(t)$ is the time-dependent probability of a transition from state n_o to state n.

The energy levels of the initial and final states often form a continuous spectrum. Let us examine in greater detail the case (which is of physical interest) when the energy of the final state is part of the continuous spectrum. We shall assume that the variables n of the final state consist of E (the energy) and of a certain set of variables (some of them may be continuous as well) which we shall denote by the index u. Since by assumption E varies continuously we can introduce a number of states with fixed values u in the energy range dE

$$dZ = \eta_u(E)dE \quad . \tag{1.7.11}$$

Here $\eta_u(E)$ is the energy density of states defined by the relation (1.7.11).

In the case of continuous (or quasi-continuous) energy spectrum, the finite transition probability can be obtained by the summation (integration) over a finite interval of energies. However, we can perform integration over all energies, since, as we will see, only energies E close to the initial energy E_o give essential contribution to the transition probability. Thus the required probability becomes

$$W_{n_o u}(t) = \sum_{\substack{E_n \\ n \neq n_o}} \rho_{nn}(t)$$

$$= \int_{E_{min}}^{E_{max}} dE\eta_u(E)2|\langle E_o, n_o | V | E, u\rangle|^2 (E - E_o)^{-2}\left[1 - \cos\left((E_o - E)t/\hbar\right)\right], \tag{1.7.12}$$

where E_{min} and E_{max} are minimal and maximal energies of the energy spectrum.

Now, under certain limitations, this expression is approximately proportional to time, and then we are able to speak about the transition probability per unit time.

Again, because of the importance of this notion, "transition probability per unit time", and frequent misunderstanding of its limitations, we will try to analyze them in detail.

Let us introduce a new variable of integration

$$y = \frac{1}{2\hbar}(E - E_o)t, \qquad E = E_o + 2\hbar y / t, \qquad dy = \frac{t}{2\hbar}dE;$$

then expression (1.7.12) may be rewritten as

$$W_{n_o u} = \frac{2\pi}{\hbar}t \int_{-(E_o - E_{min})\frac{t}{2\hbar}}^{(E_{max} - E_o)\frac{t}{2\hbar}} dy f(E_o + 2\hbar y / t) \frac{\sin^2 y}{\pi y^2} \quad , \tag{1.7.13}$$

where

$$f(E) = |< E_o, n_o \mid V \mid E, u >|^2 \ \eta_u(E) \quad . \tag{1.7.14}$$

Now, let $\Delta E = \hbar\tilde{\omega}$ denote the range in which function f(E) changes only slightly when E changes by amount δE which is small compared with $\hbar\tilde{\omega}$, $f(E_o + \delta E) \approx f(E_o)$. In other words, $\hbar\tilde{\omega}$ is the characteristic scale of the variation of E near E_o. Let us suppose that time t is so large that the following limitations are satisfied

$$t >> \tilde{\omega}^{-1}, \ \ \hbar(E_o - E_{min})^{-1}, \ \ \hbar(E_{max} - E_o)^{-1} \quad . \tag{1.7.15}$$

Now we designate $(\omega^*)^{-1}$ a maximal value among three terms in the right hand side of (1.7.15). Then condition (1.7.15) can be written in the form

$$t >> (\omega^*)^{-1} \quad . \tag{1.7.16}$$

We notice further that function $\sin^2 y/\pi y^2$ gives its main contribution to the integral (1.7.13) in the region $-\pi < y < \pi$ (i.e., for values $y^2 \sim 1$). By using this fact and the inequalities (1.7.15) and (1.7.16) we get

$$W_{n_o u} = \frac{2\pi}{\hbar} t \, |< E_o, n_o \mid V \mid E_o, u >|^2 \ \eta_u(E_o) = w_{n_o u} t \quad . \tag{1.7.17}$$

Thus, under condition (1.7.16), i.e. for large enough t, the transition probability is proportional to time t and it is possible to introduce the transition probability per unit time $w_{n_o u}$ (which itself does not depend on time). Of course the quantities determining conditions (1.7.15) and (1.7.16) depend on the actual properties of the system under investigation.

The applicability of the expression (1.7.17) depends not only on the condition (1.7.16); a condition of the perturbation theory must also be satisfied

$$w_{n_o u} t \ll 1 \quad . \tag{1.7.18}$$

In order to satisfy both conditions (1.7.16) and (1.7.18) it is necessary

$$\omega^* \gg w_{n_o u} \quad . \tag{1.7.19}$$

In this case there is such an interval of time that both conditions (1.7.16) and (1.7.18) are satisfied

$$(\omega^*)^{-1} \ll t \ll w_{n_o u}^{-1} \quad . \tag{1.7.20}$$

If condition (1.7.19) is not satisfied then there is no time-independent transition probability per unit time. It should be mentioned also that for small values of time

$$t \ll (\omega^*)^{-1} \quad , \tag{1.7.21}$$

there is no time-independent transition probability per unit time. In this case, according to (1.7.12) the transition probability is proportional to t^2.

Very frequently the transition probability per unit time is expressed in the form

$$w_{n_o u} = \frac{2\pi}{\hbar} | V_{n_o u} |^2 \, \delta(E - E_o) \quad . \tag{1.7.22}$$

It should, however, be remembered that this expression has meaning only after integration over the energy E, using Eq. (1.7.11). Then it obtains the form (1.7.17)

$$w_{n_o u} = \frac{2\pi}{\hbar} |\langle E_o, n_o | V | E_o, u \rangle|^2 \, \eta_u(E_o) \quad . \tag{1.7.23}$$

Probability (per unit time) of transition from the initial state $\langle E_o, n_o|$ to all other states $|E,u\rangle$ can be presented in the alternative and widely used form

$$w(n_o) = \hbar^{-2} \int_{-\infty}^{\infty} \langle n_o | V(\tau) V(0) | n_o \rangle \, d\tau \quad . \tag{1.7.24}$$

It is easy to show the validity of this expression by using rule (1.2.7) for a matrix element of a product of matrices

$$w(n_o) = \hbar^{-2} \int_{-\infty}^{\infty} \sum_u | V_{n_o u} |^2 \, e^{\frac{i}{\hbar}(E_{n_o} - E_u)\tau} d\tau =$$

$$=\frac{2\pi}{\hbar}\sum_{u}|V_{n_0u}|^2\,\delta(E_{n_0}-E_u)\ . \tag{1.7.25}$$

The probability of transition (per unit time) between two states characterized by discrete (m and n) and continuous indices (including energy) and averaged over the initial state u is

$$w_{nm}=\hbar^{-2}\int_{-\infty}^{\infty}<V_{nm}(t)V_{mn}(0)>e^{i\omega_{nm}\tau}d\tau\ . \tag{1.7.26}$$

Here

$$<V_{nm}(\tau)V_{mn}(0)>=\sum_{uu'}P_uV_{nu;mu'}V_{mu';nu}e^{\frac{i}{\hbar}(E_u-E_{u'})\tau}\ , \tag{1.7.27}$$

where P_u is the distribution probability of states |nu> (with fixed n).

In the case when we are dealing with the quasi-continuum, characterized by some mean energy interval δE between adjacent energy levels, the transition from the summation to integration in Eq. (1.7.12) is possible if

$$t \ll \hbar/\delta E\ . \tag{1.7.28}$$

This means that δE should satisfy the condition

$$\omega^* \gg w_{n_0u} \gg \delta E/\hbar\ , \tag{1.7.29}$$

since $w^{-1}_{n_0u}$ is the characteristic time of the problem.

1.8 Continuous spectrum of energies and irreversibility

It will be shown in the next chapter that transition probabilities per unit time characterize a certain class of irreversible processes (Markovian approximation). It has been shown that the derivation of formula (1.7.23) for the probability per unit time depends on the existence of the continuum of energies. Much more general statement can be proven. The irreversible behavior of the system is possible only for systems having continuous spectrum of energies (Golden and Longuet-Higgins, [5]: Fain [6]).

For the reversible process the mean value of a certain quantity A: <A(t)> has limit $t\to\infty$ only if <A(t)> does not depend on time. In the reversible process a system return to the initial state many times and, therefore, a variable quantity <A(t)> does not have limit at $t\to\infty$. On the other hand, for the irreversible process <A(t)> has the limit $t\to\infty$ in a general case.

The mean value of the quantity A at the time t is

$$< A(t) >= \mathrm{Tr}(\rho(t)A) \ . \tag{1.8.1}$$

The general solution of the von Neumann equation (1.1.4) can be written in the form

$$\rho(t) = e^{-iHt/\hbar}\rho(0)e^{iHt/\hbar} \ . \tag{1.8.2}$$

Therefore, the mean value of <A(t)> has the form

$$< A(t) >= \sum_{nmuu'} \rho_{nu;mu'}(0)A_{mu';nu}e^{-i\omega_{nm}t} = \int_{-\infty}^{\infty} g(\omega)e^{i\omega t}d\omega \ , \tag{1.8.3}$$

where n,m are indices denoting the energy levels, $\omega_{nm} = (E_n\text{-}E_m)/\hbar$; u,u' are other quantum numbers; and the quantity $g(\omega)$, as can easily be checked, has the form

$$g(\omega) = \sum_{uu'nm} \rho_{nu;mu'}(0)A_{mu';nu}\delta\left(\frac{E_n - E_m}{\hbar} + \omega\right) \ . \tag{1.8.4}$$

If the system has a discrete energy spectrum, $g(\omega)$ is a discontinuous function, <A(t)> is equal to the discrete sum of the harmonic functions and has no limit when $t \to \infty$. The quantum recurrence theorem (the quantum analog of the Poincaré theorem) states that the quantum system gets arbitrarily close to the initial state at arbitrarily large times [7-9].

For the system with a continuous spectrum of energies (the Poincaré recurrence time tends to infinity) <A(t)> will have a limit when $t \to \infty$, if $g(\omega)$ has the form

$$g(\omega) = G\delta(\omega) + h(\omega) \ , \tag{1.8.5}$$

where $h(\omega)$ has no δ-type singularities and is absolutely integrable in the range $[-\infty,+\infty]$. Then on the basis of the Lebesque-Riemann theorem

$$\lim_{t\to\infty} < A(t) >= \lim_{t\to\infty}\left\{G + \int_{-\infty}^{\infty} h(\omega)e^{-i\omega t}d\omega\right\} = G \ . \tag{1.8.6}$$

Thus it has been proven that irreversible behavior of quantum systems is connected with the continuous energy spectrum of the system.

CHAPTER II

QUANTUM THEORY OF RELAXATION PROCESSES

If a quantum system is closed, the time evolution of its state can be described by the von Neumann equation (1.1.4). However, we more often meet the case in which the system of interest to us is in contact with its surroundings. In particular, the system may be in contact with a thermal bath. The problem is how to describe time behavior of this kind of non-closed ("open") system. It is obvious that the solution is of major significance to various branches of physics. When we are interested in the steady state of the system (without external forces) the solution is well known. According to the basic principles of statistical physics, a system in contact with a constant temperature bath (with which it interacts weakly), and which is in a steady state (state of thermal equilibrium) can be described by the density matrix

$$\rho = \exp\left(-\frac{H}{k_B T}\right) \Big/ \mathrm{Tr} \exp\left(-\frac{H}{k_B T}\right) ,$$

where T is a temperature of the bath and H is the Hamiltonian of the "open" system we are interested in. (Only in the case of weak coupling of the system with the thermal bath we can (approximately) define the Hamiltonian H.)

Now let the system be in a non-equilibrium state. In the course of time this system will approach an equilibrium state. The processes which occur during this time are called relaxation processes. Generalizing, we shall give the name of relaxation processes to any transient process, in particular those in which a system approaches its stationary state (in the presence of an external force). Relaxation processes play an important part in physics. The following examples may be given of relaxation processes that are met in practice: the spontaneous emission in free space, the decay of an electromagnetic in a lossy resonator, relaxation of a spin system due to the interaction with the crystal lattice (spin-lattice relaxation), etc. In all these examples the relaxation takes place as a result of interaction with systems which have, in the limit, a continuous energy spectrum (see section 1.8). It will be convenient to use the following terminology. We shall call the system we are interested in a *dynamic system* and its surroundings a *dissipative system* (it is a source of the energy dissipation). It is understood here that the dynamic system (or subsystem) has usually a finite number of degrees of freedom and discrete energy levels, whilst the dissipative system has an infinite (in the limit) number of degrees of freedom and a continuous energy spectrum.

The dynamic and dissipative subsystems, interacting with each other, together form a closed system. The behavior of this closed system can be described by the von Neumann equation, using the density matrix of the whole system $\rho_{m\alpha;n\beta}$, where the Latin letters are the discrete indices of the dynamic subsystem and the Greek letters are indices of the dissipative system (at least part of them, and energy among them, run through a continuous series of values). When we are interested in the behavior of the dynamic subsystem we need to know only that part of the density matrix $\rho_{m\alpha;n\alpha}$ which is diagonal in the indices α, or its trace in the indices α

$$\sigma_{mn} = \sum_{\alpha} \rho_{m\alpha;n\alpha} \quad .$$

This chapter is supposed to present a general theoretical framework for the description of various irreversible relaxation processes. (We will not consider the systems exposed to the external time-varied forces, or fields.)

2.1 Exact equations describing temporal behavior of interacting dissipative and dynamic systems

In very general terms the theory of rate processes may be formulated as follows. We are dealing with a dynamic subsystem (a spin system, an impurity in a solid, local vibrations, etc.) interacting with the dissipative system. We will be interested in the behavior of the dynamic subsystem (or generally, in the behavior of the relevant part of the system). Generally speaking the behavior of the subsystem is determined by the von Neumann equation for the whole system. The question arises whether it is possible to find out the equations determining the temporal behavior of the subsystem only. The general answer to this question is negative. The density matrix of the subsystem at the moment t, $\sigma(t)$, is determined not only by the initial condition $\sigma(0)$, but also by the initial condition $\rho(0)$ of the density matrix of the whole system. However, in some special cases, under special physical conditions it is possible to perform an approximate reduction of the equations for the subsystem only. This will be done in forthcoming sections. The aim of this section is to provide a general framework for the description of interacting dynamic and dissipative systems. Such a framework can be established without using any approximation. The approximate equations will be derived in the next sections.

The behavior of a closed system containing dynamic and dissipative subsystems can be described by the von Neumann equation

$$i\hbar\frac{\partial\rho}{\partial t} = [H,\rho] \quad , \tag{2.1.1}$$

where H is the Hamiltonian of the closed system. This Hamiltonian may be presented in the form

$$H = \mathrm{E} + \mathrm{F} + \mathrm{V} + \mathrm{G} = \mathrm{H_o} + \mathrm{V} + \mathrm{G}\ , \tag{2.1.2}$$

where E is the Hamiltonian of the dynamic system with the eigenvalues E_n and eigenfunctions $|n>$; F is the Hamiltonian of the dissipative system with eigenvalues F_α and eigenfunctions $|\alpha>$; V is the interaction energy between dynamic and dissipative systems with matrix elements $V_{m\alpha;n\beta}$, G is the perturbation energy in the dissipative system (leading to transitions between various states $|\alpha>$ and $|\beta>$), it has matrix elements $G_{\alpha\beta}$.

To proceed further, we will use a procedure invented by Zwanzig [10,11] and Nakajima [12]. It is also called the Zwanzig formalism. Our final goal is to describe the time behavior of the dynamic system (or dissipative system). The density matrix of the whole system in the representation of the unperturbed Hamiltonian

$$H_o = E + F \quad , \tag{2.1.3}$$

has matrix elements

$$\rho_{m\alpha;n\beta} \quad .$$

The behavior of the dynamic system may be described by the density matrix

$$\sigma_{mn} = \sum_{\alpha} \rho_{m\alpha;n\alpha} \, , \qquad \sigma = \mathrm{Tr}_b \rho \quad , \tag{2.1.4}$$

where Tr_b is the trace operation over the bath variables α.

Now we try to derive the equations determining the behavior of the dynamic system, starting from Eq. (2.1.1), which is valid for the whole system. This Eq. (2.1.1) may be presented in the form

$$\frac{\partial \rho}{\partial t} = -iL\rho \quad , \tag{2.1.5}$$

where L is the Liouville operator of the system. The Liouville operator is a superoperator or supermatrix. While the usual operator A acts on the wave function ψ transforming it to another function φ

$$A\psi = \varphi \quad , \tag{2.1.6}$$

the superoperator acts on the density matrix ρ transforming it into another matrix.

The Liouville operator L is defined as

$$\begin{aligned} L\rho &= \frac{1}{\hbar}[H,\rho] = \frac{i}{\hbar}[H_o,\rho] + \frac{i}{\hbar}[V,\rho] + \frac{i}{\hbar}[G,\rho] \\ &\equiv \left(L_o + L^V + L^G\right)\rho \equiv \left(L_o + L_1\right)\rho \quad . \end{aligned} \tag{2.1.7}$$

In general we will call the operators acting on the density matrix the Liouville type operators. According to the definition

$$(L\rho)_{MN} = \sum_{M'N'} L_{MN}^{M'N'} \rho_{M'N'} \quad , \tag{2.1.8}$$

where M,N are a set of indices of the system (particularly m,α; n,β).
According to Eq. (2.1.5) and (2.1.1)

$$L_{MN}^{M'N'} = \frac{1}{\hbar}(H_{MM'}\delta_{NN'} - H_{N'N}\delta_{M'M}) , \qquad (2.1.9)$$

For the arbitrary supermatrices the multiplication rule is held

$$(AB)_{MN}^{M'N'} = \sum_{RS} A_{MN}^{RS} B_{RS}^{M'N'} , \qquad (2.1.10)$$

As the first step we will derive the equation for the density matrix $\rho_{m\alpha;n\alpha}$ diagonal in the indices of the dissipative system. The reduced density matrix σ is then obtained by the summation over α: Eq. (2.1.4). Thus we need to project general density matrix $\rho_{m\alpha;n\beta}$ on the partial density matrix $\rho_{m\alpha;n\alpha}$. The projection operator D is defined as

$$D\rho = \rho_1; \qquad (\rho_1)_{m\alpha;n\beta} = \rho_{m\alpha;n\alpha}\delta_{\alpha\beta} , \qquad (2.1.11)$$

while the total density matrix may be presented as

$$\rho = \rho_1 + \rho_2 \equiv D\rho + (1\text{-}D)\rho . \qquad (2.1.12)$$

Here ρ_2=(1-D)ρ has only off-diagonal matrix elements over Greek indices α, β. Operator D is defined as

$$D_{m\alpha;n\beta}^{m'\alpha';n'\beta'} = \delta_{\alpha\beta}\delta_{mm'}\delta_{\alpha\alpha'}\delta_{nn'}\delta_{\beta\beta'} . \qquad (2.1.13)$$

Equation (2.1.5) can be rewritten in the form

$$\frac{\partial\rho_1}{\partial t} = -iDL(\rho_1+\rho_2); \qquad \frac{\partial\rho_2}{\partial t} = -i(1-D)\,L(\rho_1+\rho_2) . \qquad (2.1.14)$$

The latter equation may also be presented in the form

$$\frac{\partial\rho_2}{\partial t} + i(1-D)L\rho_2 = -i(1\text{-}D)L\rho_1 . \qquad (2.1.15)$$

The solution of this equation may be presented as a sum of the general solution of the equation

$$\frac{\partial\rho_2}{\partial t} + i\,(1\text{-}D)L\rho_2 = 0 , \qquad (2.1.16)$$

and the specific solution of equation (2.1.15)

$$\rho_2 = \exp[-i(t-t_o)(1-D)L]\rho_2(t_o) - i\int_0^{t-t_o} d\tau \exp[-i\tau(1-D)L](1-D)L\rho_1(t-\tau)\,. \quad (2.1.17)$$

Substituting this solution into the right hand side of the first equation (2.1.14) we obtain the following equation for the density matrix ρ_1

$$\frac{\partial \rho_1}{\partial t} = -iDL\exp[-i(t-t_o)(1-D)L]\rho_2(t_o) - iDL\rho_1 - \int_0^{t-t_o} K(\tau)\rho_1(t-\tau)d\tau \ , \quad (2.1.18)$$

where

$$K(\tau) = DL\exp[-i\tau(1-D)L](1-D)L\ . \quad (2.1.19)$$

Two comments have to be made. First, these equations are <u>exact</u>. They are the result of identical transformations of the exact equations (2.1.1) or (2.1.5). Without any additional assumptions we cannot derive the equation for the matrix ρ_1 ($\rho_{m\alpha;n\alpha}$) only. Eq. (2.1.18) contains the off-diagonal part ρ_2 ($\rho_{m\alpha;n\beta}$, $\alpha \neq \beta$) at some initial moment of time $t = t_o$. Second, Eq. (2.1.18) describes not only $\rho_{m\alpha;n\alpha}$, but any other possible division of density matrix ρ by some projection operator D, which does not coincide with (2.1.13). For example, we may consider another projection operator

$$\tilde{D}^{m'\alpha';n'\beta'}_{m\alpha;n\beta} = \delta_{\alpha\beta}\delta_{mm'}\delta_{\alpha\alpha'}\delta_{nn'}\delta_{\beta\beta'}\delta_{mn} \ , \quad (2.1.20)$$

which projects ρ onto density matrices diagonal over all indices (including m and n), $\rho_{m\alpha;m\alpha}$.

The main result of this section is equation (2.1.18). This exact equation may serve as a starting point for various approximations. They will be discussed in next sections.

2.2 Relaxation of the dissipative system. Markovian approximation

Our final goal is to describe the time behavior of the dynamic system [density matrix σ, (2.1.4)]. However, we will first consider the relaxation process in the dissipative system, provided there is no interaction with the dynamic system ($V = 0$). In this case the Hamiltonian of the dissipative system equals

$$H = F + G\ . \quad (2.2.1)$$

For example the role of F may play the Hamiltonian of phonons, and the role of interaction G may play the anharmonic terms of the total Hamiltonian. Density matrix ρ of the dissipative system has matrix elements $\rho_{\alpha\beta}$, where quantum numbers α, β describe eigenstates of the Hamiltonian F. This Hamiltonian is diagonal in the $|\alpha\rangle$-representation

$$F_{\alpha\beta} = F_\alpha \delta_{\alpha\beta} \quad . \tag{2.2.2}$$

We would try to derive from the basic quantum-mechanical equations (2.1.18) – (2.1.19) the master equation for the diagonal elements of the density matrix ρ:

$$\dot{\rho}_{\alpha\alpha} = -\sum_\beta (w_{\alpha\beta}\rho_{\alpha\alpha} - w_{\beta\alpha}\rho_{\beta\beta}) \quad . \tag{2.2.3}$$

We remind that diagonal matrix elements $\rho_{\alpha\alpha}$ give the probability to find the system in the state α. Quantities $w_{\alpha\beta}$ are transition probabilities per unit time (from state α to state β). This type of equation was first established by Landau [13], Pauli [14] and Bloch [15]. Pauli, in his derivation, used the so-called repeated Random Phase Assumption (RPA). It means that such an equation has been derived for the small interval of time Δt, assuming that at the beginning of the interval the off-diagonal elements of the density matrix are zero

$$\rho_{\alpha\beta}(t_o) = 0 \; ; \qquad \alpha \neq \beta \; , \tag{2.2.4}$$

which means that phases distributed at random [the lack of coherence $g_{\alpha\beta} = 0$, (1.6.14)]. Then, as it is clear from the derivation of transition probability per unit time, formula (1.7.23), that for the small enough time interval Δt satisfying (1.7.18), one can derive the master equation. If at the end of this period Δt one makes again the same random phase assumption (RPA) it is possible to establish equation (2.2.3) for an arbitrary moment of time. Of course, such repeated RPA has no direct justification. Van Hove [16] first derived the master equation without repeated RPA, using instead of it the RPA only at the initial moment $t = 0$.

We will not repeat here the rather complicated derivation of van Hove and will use, instead, exact equation (2.1.18) as a starting point for derivation of the approximate equation (2.2.3). The projection operator *D* in our case transforms density matrix $\rho_{\alpha\beta}$ into its diagonal part. We designate this operator by P [to distinguish it from *D*, (2.1.13)]

$$\begin{aligned} &\rho = \rho_1 + \rho_2 \; ; \qquad \rho_1 = P\rho , \qquad \rho_2 = (1-P)\rho \; , \\ &(\rho_1)_{\alpha\beta} = \delta_{\alpha\beta}\rho_{\alpha\alpha} \; ; \qquad (\rho_2)_{\alpha\beta} = (1-\delta_{\alpha\beta})\rho_{\alpha\beta} \quad . \end{aligned} \tag{2.2.5}$$

The equation for diagonal elements of the density matrix ρ_1 obtains the form

$$\frac{\partial \rho_1}{\partial t} = -iPL\exp[i(t-t_o)(1-P)L]\rho_2(t_o) - iPL\rho_1 - \int_0^{t-t_o} d\tau K(\tau)\rho_1(t-\tau) \; , \tag{2.2.6}$$

where

$$K(\tau) = PL\exp[-i\tau(1-P)L](1-P)L \quad . \tag{2.2.7}$$

Thus, in the general case, no closed equation is obtained for $\rho_1(\rho_{\alpha\alpha})$, since the behavior of ρ_1 at the time t is determined not only by ρ_1 at time $t = t_o$, but also by $\rho_2(t_o)$.

Now we will show that if we assume $\rho_2(t_o) = 0$, i.e. we make RPA at the initial moment of time, then the equation (2.2.6) may be reduced to the master equation (2.2.3) provided certain approximations are performed. Taking into account the assumption (2.24), $\rho_2(t_o) = 0$, we rewrite equation (2.26) in the form

$$\frac{\partial \rho_{\alpha\alpha}}{\partial t} = -i\sum_{\beta}(PL)_{\alpha\alpha}^{\beta\beta}\rho_{\beta\beta} - \sum_{\beta}\int_0^{t-t_o} K_{\alpha\alpha}^{\beta\beta}(\tau)\rho_{\beta\beta}(t-\tau)d\tau \ . \tag{2.2.8}$$

According to Eqs. (2.2.5) operator P has the form

$$P_{\alpha\beta}^{\alpha'\beta'} = \delta_{\alpha\beta}\delta_{\alpha\alpha'}\delta_{\beta\beta'} \ , \tag{2.2.9}$$

while operator L, according to Eq. (2.1.9) has the form

$$L_{\alpha\beta}^{\alpha'\beta'} = \frac{1}{\hbar}(H_{\alpha\alpha'}\delta_{\beta\beta'} - H_{\beta'\beta}\delta_{\alpha\alpha'}) \ . \tag{2.2.10}$$

Taking into account Eqs. (2.2.1) and (2.2.2), superoperator L can be divided into two parts

$$L = L_0 + L_1 \ , \tag{2.2.11}$$

where

$$(L_0)_{\alpha\beta}^{\alpha'\beta'} = \omega_{\alpha\beta}\delta_{\alpha\alpha'}\delta_{\beta\beta'} \ ; \qquad \omega_{\alpha\beta} = \frac{1}{\hbar}(F_\alpha - F_\beta) \ , \tag{2.2.12}$$

$$(L_1)_{\alpha\beta}^{\alpha'\beta'} = \frac{1}{\hbar}(G_{\alpha\alpha'}\delta_{\beta\beta'} - G_{\beta'\beta}\delta_{\alpha\alpha'}) \ . \tag{2.2.13}$$

From (2.2.10), (2.1.10) and (2.2.9) one gets

$$(PL)_{\alpha\alpha}^{\beta\beta} = L_{\alpha\alpha}^{\beta\beta} = \hbar^{-1}(H_{\alpha\beta}\delta_{\alpha\beta} - H_{\beta\alpha}\delta_{\beta\alpha}) = 0 \ . \tag{2.2.14}$$

It is easy to show that $L_0P = PL_0$ and therefore the kernel K in Eq. (2.2.6) can be rewritten as

$$K = PL_1 \exp[-i\tau(1-P)L](1-P)L_1 \ , \tag{2.2.15}$$

and the equation for $\rho_{\alpha\alpha}$ obtains the form

$$\frac{\partial \rho_{\alpha\alpha}}{\partial t} = -\sum_{\beta} \int_{0}^{t-t_0} K_{\alpha\alpha}^{\beta\beta}(\tau)\rho_{\beta\beta}(t-\tau)d\tau \ . \tag{2.2.16}$$

This equation with kernel (2.2.15) is still exact, provided $\rho_2(t_0) = 0$. To get master equation (2.2.3) we have to perform two approximations: the Born approximation and the Markovian approximation. In the Born approximation we assume the smallness of the perturbation G and calculate the matrix K up to the terms G^2. Therefore we put in the exponent (2.2.15) instead of L its approximate value L_0

$$K = PL_1 \exp[-i\tau(1-P)L_0](1-P)L_1 \ . \tag{2.2.17}$$

In this approximation the explicit expression for the matrix element of K takes the form

$$K_{\alpha\alpha}^{\beta\beta}(\tau) = \hbar^{-2} \sum_{\alpha'\neq\alpha} |G_{\alpha\alpha'}|^2 \left(e^{i\omega_{\alpha\alpha'}\tau} + e^{-i\omega_{\alpha\alpha'}\tau}\right)\delta_{\alpha\beta}$$
$$-\hbar^{-2} |G_{\alpha\beta}|^2 \left(e^{i\omega_{\alpha\beta}\tau} + e^{-i\omega_{\alpha\beta}\tau}\right)(1-\delta_{\alpha\beta}) \ . \tag{2.2.18}$$

Thus the equation for the diagonal elements of the density matrix takes the form

$$\frac{\partial \rho_{\alpha\alpha}}{\partial t} = -\sum_{\beta} \int_{0}^{t-t_0} d\tau R_{\alpha\beta}(\tau)\left[\rho_{\alpha\alpha}(t-\tau) - \rho_{\beta\beta}(t-\tau)\right] , \tag{2.2.19}$$

where

$$R_{\alpha\beta}(\tau) = \hbar^{-2} |G_{\alpha\beta}|^2 \left(e^{i\omega_{\alpha\beta}\tau} + e^{-i\omega_{\alpha\beta}\tau}\right) . \tag{2.2.20}$$

These equations are not master equations (2.2.3). In the master equations (2.2.3) the derivative of the probability $\rho_{\alpha\alpha}(t)$ of finding a system in the state α at time t is determined by the probabilities $\rho_{\beta\beta}(t)$ at the same moment and do not depend on the prehistory of the system. In other words the system does not have the memory. As opposed to this, equation (2.2.19) does have memory. The derivative of $\rho_{\alpha\alpha}(t)$ depends on the history of the system. The equation (2.2.19) contains memory about the state of the system in all previous moments of time. The master equations (2.2.3) can be obtained from Eqs. (2.2.19) in so-called Markovian approximation. In this approximation one can expand quantities $\rho_{\alpha\alpha}(t-\tau)$ over the delay time τ

$$\rho_{\alpha\alpha}(t-\tau) = \rho_{\alpha\alpha}(t) - \tau\dot{\rho}_{\alpha\alpha}(t) + \ldots , \tag{2.2.21}$$

and retain only the first term $\rho_{\alpha\alpha}(t)$ in the right hand side of equation (2.2.19). Such an approximation is justified provided second and higher order terms may be neglected.

Let us assume that sums over functions $R_{\alpha\beta}(t)$ in (2.2.19) are decaying fast enough, τ_c being the characteristic decay time. It is easy to see that one can neglect the second term in the right hand side of equation (2.2.21) (as well as higher order terms), provided

$$\tau_c \ll T_{diss} \ , \tag{2.2.22}$$

where T_{diss} is the characteristic relaxation time of the dissipative system which can be estimated as

$$T_{diss}^{-1} \sim \sum_{\beta} \int_0^{t-t_0} d\tau R_{\alpha\beta}(\tau) \ . \tag{2.2.23}$$

Thus equation (2.2.19) can be approximated as

$$\frac{\partial \rho_{\alpha\alpha}}{\partial t} = -\sum_{\beta} w_{\alpha\beta}(t)(\rho_{\alpha\alpha} - \rho_{\beta\beta}) \ , \tag{2.2.24}$$

where $w_{\alpha\beta}(t)$ are time-dependent transition probabilities per unit time

$$w_{\alpha\beta}(t) = \int_0^{t-t_0} d\tau R_{\alpha\beta}(\tau) \ . \tag{2.2.25}$$

These coefficients are proportional to time $t-t_0$, when $t-t_0 \ll \tau_c$; while at

$$t - t_0 \gg \tau_c \ , \tag{2.2.26}$$

the transition probabilities per unit time do not depend on time. They equal

$$w_{\alpha\beta} = w_{\beta\alpha} = \int_0^{\infty} d\tau R_{\alpha\beta}(\tau) = \frac{2\pi}{\hbar} | G_{\alpha\beta} |^2 (F_\alpha - F_\beta) \ . \tag{2.2.27}$$

In this case we get the master equation (2.2.3) with $w_{\alpha\beta} = w_{\beta\alpha}$

$$\frac{\partial \rho_{\alpha\alpha}}{\partial t} = -\sum_{\beta} w_{\alpha\beta}(\rho_{\alpha\alpha} - \rho_{\beta\beta}) \ . \tag{2.2.28}$$

It is worthwhile to mention that conditions (2.2.22) and (2.2.26) coincide with conditions (1.7.19) and (1.7.16) respectively. It is easy to identify $(\omega^*)^{-1}$ with τ_c, since the sums over $R_{\alpha\beta}(t)$ in (2.2.19) are, as a matter of fact, the Fourier integrals with the characteristic frequency range equal to ω^*.

The irreversible behavior described by equation (2.2.24) or (2.2.28) has been derived from the first principle, i.e. from exact equation (2.2.6) using the Born and the Markovian approximation. However a specific initial condition has been assumed

$$\rho_2(t_0) = 0 \quad , \tag{2.2.29}$$

corresponding to RPA at the initial time $t = t_0$. Of course, such an assumption is not a very satisfactory one. The moment of time $t = t_0$ is quite arbitrary, and at any time $t \neq t_0$, $\rho_2(t)$ is already not zero [see e.g. (2.1.17)]. However, the sums which are inherent in the first term in the right hand side of Eq. (2.2.6)

$$-iPL\exp[it(t-t_0)(1-P)L]\rho_2(t_0) \quad , \tag{2.2.30}$$

are, as a matter of fact, the Fourier integrals (due to the assumption of continuous energy spectrum of the dissipative system). These Fourier integrals are characterized by certain frequency regions $\Delta\omega$, which in many cases have orders of magnitudes $\Delta\omega$ satisfying

$$\Delta\omega \gg T_{diss}^{-1} \quad , \tag{2.2.31}$$

This condition is similar to (2.2.22). In this case term (2.2.30) decays during the time $\tau_c = \Delta\omega^{-1}$, which is much smaller than the relaxation time.

It is expedient to mention that van Hove [16] has derived the master equation (2.2.3) both for initial conditions $\rho_2(t_0) = 0$ and for initial conditions satisfying (2.2.31). We have to recognize that in all the above approximations and derivations a certain unsatisfactory element is present. It does not matter which initial conditions $\rho_2(t_0)$ we assume. There is a certain time interval

$$t_0 < t \lesssim \tau_c \quad , \tag{2.2.32}$$

where the conventional equations with time-independent coefficients are not valid. [In the region (2.2.32) equations (2.2.24) with time-dependent rate coefficients are valid.] It may mean that one cannot use arbitrary initial conditions using conventional master equations. In the region close to the initial conditions these master equations are not valid. We will discuss this point more in detail in section 2.6.

2.3 Equations for density matrix of dynamic systems

As has been mentioned above, in very general terms the theory of relaxation processes may be formulated as follows. We are dealing with the dynamic subsystem (an atom, a local vibration, a spin system, a certain mode of electromagnetic field in the resonator cavity) interacting with the dissipative system having an infinite (in the limit) number of degrees of freedom and a continuous energy spectrum. The question arises whether it is possible to find equations determining the temporal behavior of the dynamic system only. The density matrix of the dynamic subsystem at the moment t, $\sigma(t)$, is determined by the equations for the density matrix $\rho_{m\alpha;n\alpha}$, (2.1.18). However, under special physical conditions it is possible to perform an approximate reduction to the equations for the subsystem density matrix only. The aim of this section is to show how this approximate reduction may be performed.

We will start from exact equations (2.1.18) with D defined by (2.1.13) and with the Hamiltonian (2.1.2). The Liouville operator has the form

$$L = L_0 + L^V + L^G \equiv L_0 + L_1 \quad , \tag{2.3.1}$$

where

$$\begin{aligned}(L_0)^{m'\alpha';n'\beta'}_{m\alpha;n\beta} &= \frac{1}{\hbar}(H_{0m\alpha m'\alpha'}\delta_{nn'}\delta_{\beta\beta'} - H_{0m'\beta';n\beta}\delta_{mm'}\delta_{\alpha\alpha'}) = \\ &= (\omega_{mn} + \omega_{\alpha\beta})\delta_{mm'}\delta_{\alpha\alpha'}\delta_{nn'}\delta_{\beta\beta'} \quad ,\end{aligned} \tag{2.3.2}$$

where $\hbar\omega_{mn} = E_m - E_n$, $\hbar\omega_{\alpha\beta} = F_\alpha - F_\beta$, and it is assumed that both E and F, (2.1.3), are diagonal in m and α indices ($|m\rangle$ and $|\alpha\rangle$ are eigenfunctions of E and F respectively). The Liouville operators L^V and L^G

$$(L^V)^{m'\alpha';n'\beta'}_{m\alpha';n\beta} = \frac{1}{\hbar}(V_{m\alpha;m'\alpha'}\delta_{nn'}\delta_{\beta\beta'} - V_{n'\beta';n\beta}\delta_{mm'}\delta_{\alpha\alpha'}) \quad , \tag{2.3.3}$$

$$(L^G)^{m'\alpha';n'\beta'}_{m\alpha';n\beta} = \frac{1}{\hbar}(G_{\alpha\alpha'}\delta_{mm'}\delta_{nn'}\delta_{\beta\beta'} - G_{\beta'\beta}\delta_{nn'}\delta_{mm'}\delta_{\alpha\alpha'}) \quad . \tag{2.3.4}$$

It is easy to show that the superoperator L_0 (2.3.2) commutes with D, (2.1.13)

$$DL_0 = L_0D \quad . \tag{2.3.5}$$

Therefore the kernel (2.1.19) can be rewritten as

$$K(\tau) = DL_1\exp[-i\tau(1-D)L]\,(1-D)L_1 \quad . \tag{2.3.6}$$

As in the previous section we assume the Born approximation, which means that we may approximately replace L in the exponent (2.3.6) by L_0. In this approximation superoperator $K(\tau)$ can be presented in the form

$$K(t) = K^{VV}(\tau) + K^{GG}(\tau) + K^{GV}(\tau) \quad , \tag{2.3.7}$$

where

$$K^{VV}(\tau) = DL^V\exp[-i\tau(1-D)L_0]\,(1-D)L^V \,, \tag{2.3.8}$$

$$K^{GG}(\tau) = DL^G\exp[-i\tau(1-D)L_0]\,(1-D)L^G \,,$$

$$= D\text{L}^{\text{V}}\exp[-i\tau(1-D)\text{L}_0]\,(1-D)\text{L}^{\text{G}}\ , \tag{2.3.9}$$

$$+ D\text{L}^{\text{G}}\exp[-i\tau(1-D)\text{L}_0]\,(1-D)\text{L}^{\text{V}}\ . \tag{2.3.10}$$

In more explicit form we can rewrite equation (2.1.18), with K(τ) given by (2.3.6),

$$\frac{\partial \rho_{m\alpha;n\alpha}}{\partial t} = -i \sum_{l,k;\alpha',\beta'} \{D\text{L}\exp[-i(t-t_0)(1-D)\text{L}]\}_{m\alpha;n\alpha}^{l\alpha';k\beta'} \rho_{l\alpha';k\beta'}(t_o)(1-\delta_{\alpha'\beta'})$$

$$-i\omega_{mn}\rho_{m\alpha;n\alpha} - \frac{1}{\hbar}\sum_k \left(V_{m\alpha;k\alpha}\rho_{k\alpha;n\alpha} - \rho_{m\alpha;k\alpha}V_{k\alpha;n\alpha}\right) - \sum_\beta w_{\alpha\beta}\left(\rho_{m\alpha;n\alpha} - \rho_{m\beta;n\beta}\right)$$

$$-\sum_{l,k,\beta} \int_{t_o}^{t-t_0} \left\{K^{VV}(\tau) + K^{GV}(\tau)\right\}_{m\alpha;n\alpha}^{l\beta;k\beta} \rho_{l\beta;k\beta}(t-\tau)d\tau\ . \tag{2.3.11}$$

Here the Markovian approximation for the dissipative system is assumed [the fourth term in the right hand side of Eq. (2.3.11)]. Equation (2.3.11) determines the time evolution of the density matrix $\rho_{m\alpha;n\alpha}$ (under certain initial conditions for the off-diagonal in Greek indices matrix $\rho_{m\alpha;n\beta}(t_0)$). In the general case, this equation cannot be reduced to the equation for the density matrix (2.1.4)

$$\sigma_{mn} = \sum_\alpha \rho_{m\alpha;n\alpha}\ . \tag{2.3.12}$$

However, under certain conditions such a reduction can be approximately performed [17,18]. These conditions correspond to the fast relaxation of the dissipative system in comparison with the relaxation of the dynamic subsystem. Namely, we assume that

$$T_{diss} << T_{dyn}\ , \tag{2.3.13}$$

where T_{dyn} is the characteristic relaxation time of the dynamic system. We also assume that the relaxation of the dissipative system is governed by the master equation (2.2.3) ($w_{\alpha\beta} = w_{\beta\alpha}$).

Condition (2.3.13) is equivalent to the condition

$$r = \frac{|V|^2}{|G|^2} << 1\ . \tag{2.3.14}$$

Here V and G are characteristic matrix elements of operators V and G. Conditions (2.3.13) and (2.3.14) allow us to seek the solution of the equation (2.3.11) in the form

$$\rho_{m\alpha;n\alpha}(t) = \sigma_{mn}(t)P_\alpha + \eta_{mn\alpha}; \ \sum_\alpha \eta_{mn\alpha} = 0 \ , \tag{2.3.15}$$

here P_α is the equilibrium distribution in the dissipative system. The last equation follows from the normalization condition

$$\sum_\alpha P_\alpha = 1 \ . \tag{2.3.16}$$

The term $\eta_{mn\alpha}(t)$ represents a statistical correlation between dynamic and dissipative subsystems. It is clear that if this correlation is small compared to $\sigma_{mn}(t)P_\alpha$ then we can get the equation for the density matrix $\sigma(t)$ only. Detailed conditions of smallness of $\eta_{mn\alpha}$ in the case (2.3.14), $r \ll 1$, are discussed in Appendix A to this chapter. It is clear from physical considerations that fast relaxation of the dissipative system would destroy the correlation $\eta_{mn\alpha}$ created by the interaction between the dynamic and dissipative subsystems. In this case, (2.3.14), one can use the <u>factorization approximation</u>

$$\rho_{m\alpha;n\alpha}(t) = \sigma_{mn}(t) \cdot P_\alpha \ , \tag{2.3.17}$$

which is valid at

$$t - t_0 \gg T_{dis} \ . \tag{2.3.18}$$

Expression (2.3.17) is substituted in the r.h.s. of Eq. (2.3.11).

The problem of the factorization approximation does not exist in the works of van Hove [16], Zwanzig [10,11], and Fain [20,21]. Van Hove and Zwanzig considered the relaxation dissipative system (the diagonal part of its density matrix $\rho_{\alpha\beta}$), while Fain derived equations for the diagonal in Greek indices part of the density matrix $\rho_{m\alpha;n\alpha}$. An important comment has to be made. As is clear from Appendix A, the derivation of the factorization approximation, under condition $r \ll 1$, cannot be performed in the case of the zero temperature of the thermal bath (dissipative system). The spontaneous emission and other processes taking place at the boson vacuum are considered in Chapter V. They do not need introduction of the interaction G.

Substituting the factorized density matrix, (2.3.17), into the r.h.s. of equation (2.3.11), performing the summation over indices α in both sides of Eq. (2.3.11) (the first term in the r.h.s. of Eq. (2.3.11) can be neglected provided condition (2.3.18) is satisfied, see Appendix A), we obtain the equation for the density matrix $\sigma(t)$ of the dynamic subsystem

$$\hat{\sigma} = -\frac{i}{\hbar}[E + \overline{V}, \sigma] - \int_0^{t-t_0} \left[\overline{K}^{VV}(\tau) + \overline{K}^{VG}(\tau)\right] \sigma(t-\tau)d\tau \ , \tag{2.3.19}$$

where

$$\bar{V}_{nm} = \sum_{\alpha} P_\alpha V_{m\alpha;n\alpha} \quad , \tag{2.3.20}$$

$$\bar{K}_{mn}^{m'n'}(\tau) = \sum_{\alpha,\beta} K_{m\alpha;n\alpha}^{m'\beta;n'\beta}(\tau) P_\beta \quad . \tag{2.3.21}$$

Here $\bar{V}$ and $\bar{K}$ are operators describing the dynamic subsystem, they are obtained by the averaging operators V and K of the whole system over the equilibrium distribution of the dissipative system. Equation (2.3.19) is a closed reduced dynamics equation valid for all

$$t \geq t_0 \quad ,$$

provided special initial conditions are satisfied

$$\rho_{m\alpha;n\beta}(t_0) = \sigma_{mn}(t_0) P_\alpha \delta_{\alpha\beta} \quad , \tag{2.3.22}$$

(otherwise Eq. (2.3.19) is valid for t satisfying conditions (2.3.18)).

The initial state (2.3.22) may be obtained, e.g., by imposing an ultrashort (femtosecond) laser pulse on the system. We assume that before the pulse is imposed, the whole system is in the equilibrium state. Assuming that the laser pulse does not affect the dissipative system but causes the density matrix "jump" [22] we get the initial condition (2.3.22).

In the conclusion of this section an important comment has to be made. Sometimes the factorization approximation is justified by the assertion that the deviation from the factorized density matrix is a small quantity of the order of magnitude λ, the measure of the strength of the interaction V in the Hamiltonian (2.1.2), while the interaction G is not taken into account: $T_{diss}^{-1} = \gamma \to 0$. In this case r $\to \infty$, and it can be shown that the correlation $\eta_{mn\alpha}$ (2.3.15) (if $\eta_{mn\alpha}(0) = 0$) has the form

$$\bar{\eta}_{mn\alpha} \approx -\dot{\bar{\sigma}}_{mn} P_\alpha (t - t_0) \sim -\frac{(t - t_0)}{T_{dyn}} \bar{\sigma}_{mn} P_\alpha \quad . \tag{2.3.23}$$

Thus the deviation from the factorized density matrix is of the order $\lambda^2(t\text{-}t_0)$ which is not a small quantity. On the other hand, in the case when the parameter r, (2.3.14), is small the deviation from the factorized density matrix is of the order, (2.A9)

$$\bar{\eta}_{mn\alpha} \lesssim \frac{1}{\gamma T_{dyn}} \bar{\sigma}_{mn} P_\alpha \approx \frac{T_{diss}}{T_{dyn}} \bar{\sigma}_{mn} P_\alpha << \bar{\sigma}_{mn} P_\alpha \quad . \tag{2.3.24}$$

Thus $\bar{\eta}_{mn\alpha}$ is small provided conditions (2.3.13) and (2.3.14) are satisfied.

2.4 Generalized master equations

In the explicit form the equation (2.3.19) may be written as

$$\dot{\sigma}_{mn} = -i\omega_{mn}\sigma_{mn} - \frac{1}{\hbar}\sum_k \left(\bar{V}_{mk}\sigma_{kn} - \sigma_{nk}\bar{V}_{kn}\right)$$

$$-\int_0^{t-t_0} \sum_{lk}\left[\bar{K}^{VV}(\tau) + \bar{K}^{GV}(\tau)\right]_{mn}^{lk} \sigma_{lk}(t-\tau)d\tau \ . \quad (2.4.1)$$

The explicit form of the matrix elements $\bar{K}^{VV}$, $\bar{K}^{GV}$ is

$$\hbar^2\left(\bar{K}^{VV}(\tau)\right)_{mn}^{lk} = \left[\delta_{nk}\sum_r < \tilde{V}_{nr}(\tau)\tilde{V}_{rl}(0) > e^{-i\omega_{rn}\tau} + \delta_{ml}\sum_r < \tilde{V}_{kr}(0)\tilde{V}_{rm}(\tau) > e^{-i\omega_{mr}\tau}\right]$$

$$-\left\langle \tilde{V}_{kn}(0)\tilde{V}_{ml}(\tau)\right\rangle e^{-i\omega_{ln}\tau} - \left\langle \tilde{V}_{kn}(\tau)\tilde{V}_{ml}(0)\right\rangle e^{-i\omega_{mk}\tau} \ , \quad (2.4.2)$$

and

$$\hbar^2\left(\bar{K}^{VG}(\tau)\right)_{mn}^{lk} = \delta_{nk}\left\langle [\tilde{V}_{ml}(\tau), \tilde{G}(0)]\right\rangle e^{-i\omega_{ln}\tau}$$

$$+\delta_{ml}\left\langle [\tilde{G}(0), \tilde{V}_{kn}(\tau)]\right\rangle e^{-i\omega_{mk}\tau} \ . \quad (2.4.3)$$

Here <...> means averaging over the equilibrium state of the dissipative system, dependence of the matrix elements on time is determined by the unperturbed Hamiltonian F of the dissipative system, and tilde ~ means that only off-diagonal elements of the matrix V and G are taken into account

$$\tilde{V}_{m\alpha;n\beta} = V_{m\alpha;n\beta}(1-\delta_{\alpha\beta}), \quad \tilde{G}_{\alpha\beta} = G_{\alpha\beta}(1-\delta_{\alpha\beta}) \ . \quad (2.4.4)$$

Equation (2.4.1) contains the memory. The derivative of $\sigma_{mn}(t)$ depends on the density matrix at all the previous moments of time, starting from t_0 till t. This equation may be essentially simplified in the Markovian approximation. However, the derivation now differs from that of section 2.2, Eq. (2.2.21). The matter is that $\sigma_{mn}(t)$ may have a fast change in time due to the first term in the equation (2.4.1), and therefore we cannot use an expansion of the (2.2.21) type. Presenting $\sigma_{mn}(t)$ in the form

$$\sigma_{mn}(t) = \bar{\sigma}_{mn}(t)e^{-i\omega_{mn}t} \ , \quad (2.4.5)$$

and inserting it into equation (2.4.1) we get

$$\dot{\bar{\sigma}}_{mn}(t) = -\frac{i}{\hbar}\sum_k \left(\bar{V}_{mk}\bar{\sigma}_{kn}e^{i\omega_{mk}t} - \bar{\sigma}_{mk}\bar{V}_{kn}e^{i\omega_{kn}t}\right)$$

$$-\sum_{l,k}\int_0^{t-t_0}\left[\bar{K}^{VV}(\tau)+\bar{K}^{GV}(\tau)\right]_{mn}^{lk}\bar{\sigma}_{lk}(t-\tau)e^{i\omega_{lk}\tau}e^{i(\omega_{mn}-\omega_{lk})t} \quad . \tag{2.4.6}$$

Now we can neglect memory in this equation, i.e., substitute $\bar{\sigma}_{lk}(t)$ instead of $\bar{\sigma}_{lk}(t-\tau)$ provided the characteristic correlation time τ_c of $\bar{K}(\tau)$ (the decay time of $\bar{K}(\tau)$) satisfies condition

$$\tau_c << T_{dyn}, \frac{\hbar}{|V_{nm}|} \quad . \tag{2.4.7}$$

Here T_{dyn} is the relaxation time of the dynamic subsystem. Returning to the initial representation we get the equation

$$\dot{\sigma} = -ih^{-1}\left[E + \bar{V}^{ef}(t), \sigma\right] - \Re(t)\sigma \tag{2.4.8}$$

where E is the Hamiltonian of the dynamic system

$$\bar{V}^{ef} = \bar{V} + i\hbar^{-1}\int_0^{t-t_0}\langle[V(\tau), G(0)]\rangle d\tau,$$

$$\bar{V}_{nm} = \sum_\alpha P_\alpha V_{n\alpha;m\alpha} \quad , \tag{2.4.9}$$

and time-dependent supermatrix $\Re$ is defined by the expression

$$\Re_{mn}^{lk}(t) = \int_0^{t-t_0}(\bar{K}^{VV}(\tau))_{mn}^{lk}e^{i\omega_{lk}\tau}d\tau \ . \tag{2.4.10}$$

Starting from time

$$t - t_0 >> \tau_c \quad \text{and} \quad T_{diss} \ , \tag{2.4.11}$$

the supermatrix $\Re$ and operator V^{ef} do not depend on time and have the form

$$\Re_{mn}^{lk} = \int_0^\infty (\bar{K}^{VV}(\tau))_{mn}^{lk}e^{i\omega_{lk}\tau}d\tau \ , \tag{2.4.12}$$

$$\bar{V}^{ef} = \bar{V} + i\hbar^{-1}\int_0^\infty <[V(\tau), G(0)]> d\tau \ . \tag{2.4.13}$$

In the case (2.4.11), Equation (2.4.8) is approximated by the equation with constant $\bar{V}^{ef}$ and constant supermatrix $\Re = \lim_{t \to \infty} \Re(t)$

$$\dot{\sigma} = -i\hbar^{-1}[E + \bar{V}^{ef}, \sigma] - \Re\sigma \quad . \tag{2.4.14}$$

This equation, without term $\bar{V}^{ef}$, sometimes is called the Redfield equation.

Using Eqs. (2.4.2) and (2.4.12) we can present an explicit expression for the supermatrix matrix $\Re$

$$\Re_{mn}^{m'n'} = \delta_{nn'}\sum_{m_1}\Gamma^{+}_{mm_1m_1m'} + \delta_{mm'}\sum_{m_1}\Gamma^{-}_{n'm_1m_1n} - \Gamma^{-}_{n'nmm'} - \Gamma^{+}_{n'nmm'} \quad , \tag{2.4.15}$$

where

$$\Gamma^{+}_{ljik} = \hbar^{-2}\int_0^\infty d\tau e^{-i\omega_{ik}\tau}\left\langle \tilde{V}_{lj}(\tau)\tilde{V}_{ik}(0)\right\rangle_b \quad , \tag{2.4.16}$$

$$\Gamma^{-}_{ljik} = \hbar^{-2}\int_0^\infty d\tau e^{i\omega_{lj}\tau}\left\langle \tilde{V}_{lj}(0)\tilde{V}_{ik}(\tau)\right\rangle_b \quad . \tag{2.4.17}$$

Here $< >_b$ means averaging over the dissipative system (thermal bath) variables. Assuming that the Hamiltonian F of thermal bath is diagonal, Eq. (2.2.2), the expressions for $\Gamma^{\pm}$, Eqs. (2.4.16), (2.4.17), can be presented as

$$\Gamma^{+}_{ljik} = \hbar^{-2}\sum_{\alpha,\beta}\int_0^\infty d\tau e^{-i\omega_{ik}\tau} P_\alpha V_{l\alpha;j\beta} V_{i\beta;k\alpha} e^{i\omega_{\alpha\beta}\tau}(1-\delta_{\alpha\beta})$$

$$= \hbar^{-2}\sum_{\alpha\beta} P_\alpha V_{l\alpha;j\beta} V_{i\beta;k\alpha}\left\{\pi\delta(\omega_{ki}+\omega_{\alpha\beta}) + i\frac{P}{\omega_{ki}+\omega_{\alpha\beta}}\right\}(1-\delta_{\alpha\beta}) \quad , \tag{2.4.18}$$

$$\Gamma^{-}_{lijk} = \frac{1}{\hbar^2}\sum_{\alpha\beta} P_\alpha V_{l\alpha;j\beta} V_{i\beta;k\alpha}\left\{\pi\delta(\omega_{jl}+\omega_{\alpha\beta}) + i\frac{P}{\omega_{jl}+\omega_{\alpha\beta}}\right\}(1-\delta_{\alpha\beta}) \quad . \tag{2.4.19}$$

Thus, we have shown that supermatrix $\Re(t)$ has a limit $\Re \neq 0$ when $t \to \infty$. This limit is presented by Eqs. (2.4.15) and (2.4.18), (2.4.19). It should be stressed that this conclusion has been reached using the assumption that the thermal bath energy operator F is diagonalized, its matrix has the form (2.2.2).

Of course, the interaction energy G in Eq. (2.1.2) could be absorbed into the energy of the dissipative system, with

$$F' = F + G \quad , \tag{2.4.20}$$

being the energy of the dissipative system, and the total Hamiltonian having the form

$$H = E + F' + V \ . \tag{2.4.21}$$

Using the Nakajima-Zwanzig formalism, the factorization and weak coupling approximations, one can obtain equations for σ [23-25]. There are certain difficulties in the application of such equations to the concrete cases. (1) Operator F' is off-diagonal while operator F is diagonal, (2.2.2). (2) Parameter r, (2.3.13), does not appear explicitly in the Hamiltonian (2.4.21), therefore the applicability of the factorization approximation is not explicit.

Equations (2.4.14 – 2.4.17), (2.4.12, 2.4.13) are also called generalized master equations. Equations of these types have been established in the pioneering works of Wangsness and Bloch [26]. Bloch [27,28], Redfield [29], Fano [30], and other authors [31,32,6] have subsequently given similar theories. In all these papers it is assumed that the Hamiltonian of the combined dynamic + dissipative system has the form

$$H = \mathrm{E} + \mathrm{F} + \mathrm{V} \ , \tag{2.4.22}$$

where F is a diagonal operator.

The factorization approximation, which is needed to derive the reduced dynamics equations of type (2.4.6), has been justified by the assertion that the deviation from the factorized density matrix is a small quantity of order λ^2, where λ is the parameter characterizing the interaction V. However, when G = 0, the deviation from the factorized density matrix has the form $\lambda^2(t-t_0)$, according to (2.3.24), and it is not small when $\lambda \to 0$ and $t - t_0 \to \infty$. On the other hand when the parameter r, (2.3.14), is small the reduced dynamic equations can be derived.

2.5 Time convolutionless equations. Argyres and Kelley projection operators and reduced dynamics.

In the previous sections the reduced dynamics equations for $\sigma_{mn}(t)$ were obtained via a two-stage procedure. First, we have used the operator D, (2.1.13), projecting the density matrix of the combined dynamic and dissipative system on the diagonal in the indices α, β part

$$D\rho = \rho_1; \qquad (\rho_1)_{m\alpha;n\beta} = \rho_{m\alpha;n\alpha}\delta_{\alpha\beta} \ . \tag{2.5.1}$$

This operator is inserted into the Zwanzig formalism equations, (2.1.18) and (2.1.19), yielding equations for the density matrix $\rho_{m\alpha;n\alpha}$. At the second stage the factorization approximation is applied, and the reduced dynamic equations (2.3.19), (2.4.8), and (2.4.14) are obtained, provided conditions (2.3.13) and (2.3.14) are satisfied.

In a number of works [34-37, 23-25] another one-stage projection procedure is employed. The projection operator projecting the density matrix ρ into the product of the equilibrium density matrix ρ_b of the dissipative system and time-dependent density matrix $\sigma_{mn}(t)$ of dynamic system

$$\mathrm{P}\ldots = \rho_b \mathrm{Tr}_b \ldots; \quad \mathrm{P}\rho(t) = \rho_b \sigma(t) \ , \tag{2.5.2}$$

has been suggested by Argyres and Kelley [34]. Inserting this projection operator into Zwanzig [10,11] - Nakajima [12] equations (2.1.18) we get immediately equations for $\sigma(t)$. To develop the approximation procedure we have to employ a smallness of a certain parameter to get the approximate equations for $\sigma(t)$ and verify that corrections of the next order approximation are small.

To exemplify the situation we will introduce a modification of the Nakajima – Zwanzig [10-12] formalism suggested by Chaturvedi and Shibata [35]. Of course this formalism has importance in its own. Chaturvedi and Shibata have developed a projection operator approach that leads to the time convolutionless relaxation equation. The exact meaning of the words "time convolutionless" will be clear from the explanation given below.

Let the Hamiltonian of the combined dynamic and dissipative system be presented by Eq. (2.1.2)

$$H = \mathrm{E} + \mathrm{F} + \mathrm{H}_1 \equiv \mathrm{H}_0 + \mathrm{H}_1 \ , \tag{2.5.3}$$

Where

$$\mathrm{H}_1 = \mathrm{V} + \mathrm{G} \ .$$

Transforming Eq. (2.1.5), with Hamiltonian (2.5.3), to the interaction representation

$$\rho_{int}(t) = e^{i\mathrm{H}_0 t/\hbar}\rho(t)e^{-i\mathrm{H}_0 t/\hbar} \ ; \ \mathrm{H}_0 = \mathrm{E} + \mathrm{F} \ , \tag{2.5.4}$$

one obtains the Liouville equation in the interaction representation

$$\frac{\partial \rho_{int}}{\partial t} = -i\mathrm{L}_{int}\rho_{int} \ , \tag{2.5.5}$$

where L_{int} is defined as

$$\mathrm{L}(t)\ldots = \frac{1}{\hbar}[\mathrm{H}_1(t),\ldots] \ , \tag{2.5.6}$$

$$\mathrm{H}_1(t) = e^{i\mathrm{H}_0 t/\hbar}\mathrm{H}_1 e^{-i\mathrm{H}_0 t/\hbar} \ . \tag{2.5.7}$$

Here and further on we omit the index "int", but the interaction representation is expressed by the explicit time dependence of corresponding operators and super-operators.

Using arbitrary projection operators D and Q = 1-D (which may coincide with P, (2.5.2) and Q = (1-P)) and operating on Eq. (2.5.5) with D and then Q, gives the pair of coupled equations

$$\frac{\partial D\rho(t)}{\partial t} = -iDL(t)\,[D\rho(t) + Q\rho(t)] \;, \tag{2.5.8}$$

$$\frac{\partial Q\rho(t)}{\partial t} = -iQL(t)\,[D\rho(t) + Q\rho(t)] \;. \tag{2.5.9}$$

Formally integrating the equation of motion for $Q\rho(t)$, one gets

$$Q\rho(t) = g(t,0)Q\rho(0) - i\int_0^t d\tau g(t,\tau)\, QL(\tau)D\rho(\tau) \;, \tag{2.5.10}$$

where

$$g(t,\tau) \equiv \overleftarrow{T}\exp\left[-i\int_\tau^t ds QL(s)\right] \;, \tag{2.5.11}$$

and $\overleftarrow{T}$ denotes the time-ordering operator such that time arguments increasing from right to left.

Substitution of $Q\rho(t)$, (2.5.10), into Eq. (2.5.8), yields usual Nakajama- Zwanzig [10-12], which contains time convolution of $D\rho(t)$ at earlier times. To obtain a time convolutionless form, one formally solves the equation of motion $\rho(t)$, (2.5.5), to relate $\rho(t)$ at earlier time τ to a later time t by

$$\rho(\tau) = G(t,\tau)\rho(t) \;, \tag{2.5.12}$$

where

$$G(t,\tau) = \overrightarrow{T}\exp\left[i\int_\tau^t ds L(s)\right] \;, \tag{2.5.13}$$

and $\overrightarrow{T}$ denotes the anti-time-ordering operator such that time arguments increase from left to right.

Substituting Eq. (2.5.12) into Eq. (2.5.10) one obtains

$$Q\rho(t) = g(t,0)Q\rho(0) - i\int_0^t d\tau g(t,\tau)QL(\tau)\, DG(t,\tau)(D+Q)\rho(\tau) \;. \tag{2.5.14}$$

Solving this equation for $Q\rho(t)$ one obtains

$$Q\rho(t) = \left[1-\sum(t)\right]^{-1} g(t,0)Q\rho(0) + \left[1-\sum(t)\right]^{-1}\sum(t)\, D\rho(t) \;, \tag{2.5.15}$$

where

$$\sum(t) \equiv -i\int_0^t d\tau g(t,\tau)Q(t)\,\mathcal{D}G(t,\tau) \ . \tag{2.5.16}$$

Finally, substituting $Q\rho(t)$ into Eq. (2.5.8) the convolutionless equation of motion for $\mathcal{D}\rho(t)$ is obtained, it has the form [35]

$$\frac{\partial D\rho(t)}{\partial t} = K(t)\mathcal{D}\rho(t) + I(t)Q\rho(0) \ , \tag{2.5.17}$$

where

$$K(t) = -i\mathcal{D}L(t)\left[1-\sum(t)\right]^{-1}\mathcal{D} \ , \tag{2.5.18}$$

$$I(t) = -i\mathcal{D}L(t)\left[1-\sum(t)\right]^{-1}g(t,0)Q \ . \tag{2.5.19}$$

Equation (2.5.17), as opposed to Eq. (2.1.18), does not contain time convolution of $\mathcal{D}\rho(t)$. Instead, the coefficients of the inhomogeneous linear equation for $\mathcal{D}\rho(t)$ are time-dependent.

An inhomogeneous term in the right hand side of Eq. (2.5.17) is zero for initial density matrix $Q\rho(0) = 0$, i.e. for initially uncorrelated dissipative and dynamic subsystems

$$\rho(0) = \rho_b\sigma(0) \ . \tag{2.5.20}$$

Equations (2.5.17) – (2.5.19) and (2.5.16) are exact. In the weak coupling approximation (or the Born approximation) the perturbation expansion of Σ can be performed [37]

$$\sum = \sum{}^{(1)} + \sum{}^{(2)} + \dots \tag{2.5.21}$$

where

$$\sum{}^{(1)}(t) = -i\int_0^t dt_1 QL(t_1)\mathcal{D} \ , \tag{2.5.22}$$

$$\sum{}^{(2)}(t) = -\int_0^t dt_1\int_0^t dt_2\ [QL(t_1)QL(t_2)\mathcal{D} - QL(t_2)\mathcal{D}L(t_1)] \ , \tag{2.5.23}$$

$$\sum{}^{(3)}(t) = \int_0^t dt_1\int_0^{t_1} dt_2\int_0^{t_2} dt_3\ [QL(t_1)QL(t_2)QL(t_3)\mathcal{D}$$

$$+QL(t_3)DL(t_2)L(t_1) - QL(t_2)QL(t_3)DL(t_1)\text{-}QL(t_1)QL(t_1)DL(t_2)] \; . \quad (2.5.24)$$

Explicit expressions for the first terms of K(t) are [37]

$$K^{(1)}(t) = \text{-}iDL(t)D \; , \quad (2.5.25)$$

$$K^{(2)}(t) = -\int_0^t dt_1 \; DL(t)QL(t_1)D \; , \quad (2.5.26)$$

$$K^{(3)}(t) = iDL(t)\int_0^t dt_1 \int_0^{t_1} dt_2 \; [QL(t_1)QL(t_2)D - QL(t_2)DL(t_1)D] \; . \quad (2.5.27)$$

The expressions for the first three terms $\Gamma^{(n)}$ are identical to those for $K^{(n)}$ except the final D in each term is replaced by $Q = (1\text{-}D)$.

As has been mentioned, Eq. (2.5.17) is exact and valid for the arbitrary projection operator $D = D^2$, and $Q = 1\text{-}D$. However, it is not obvious that an arbitrary projection operator D will lead to the non-diverging expansions for Σ, (2.5.21) and K (2.5.25) – (2.5.27). The conventionally accepted Hamiltonian of interacting the dynamic and dissipative subsystems does not contain the interaction energy G in (2.5.2) and (2.5.3)

$$H = E + F + V \; , \quad (2.5.28)$$

where V is the interaction energy between the dynamic and dissipative systems and F is a diagonalized energy operator. As has been shown in Section 2.4, in the case G = 0, the deviation from the factorized density matrix

$$\eta_{mn\alpha}(t) = \rho_{m\alpha;n\alpha}(t) - \sigma_{mn}(t)P_\alpha \; , \quad (2.5.29)$$

is proportional to $\lambda^2 t$ (here and further on we put $t_0 = 0$) and is not small. It explains how higher order approximation corrections $\sum^{(n)}$ and $K^{(n)}$ can be not small, though formally they are proportional to powers of λ.

Thus, substitution of the Argyres and Kelley [34] projection operator (2.5.2) into equations (2.5.27) – (2.5.18) leads to the Redfield type equations in the lowest approximation, while higher order corrections can diverge. To prove this assumption we substitute the Argyres and Kelley [34] projection operator P, (2.5.2), into equations (2.5.17) - (2.5.19) and (2.5.16) ($D = P$) and obtain a rate equation for $\sigma(t)$ that has an inhomogeneous term

$$\frac{\partial \sigma}{\partial t} = -R(t)\sigma + J(t) \; , \quad (2.5.30)$$

where the approximate equation for $\sigma(t)$ up to the second order of λ has the form

$$\frac{\partial\sigma}{\partial t} = -\frac{1}{\hbar}[\bar{V},\sigma] - R^{(2)}(t)\sigma + J^{(1)}(t) + J^{(2)}(t) \quad , \tag{2.5.31}$$

where

$$R^{(2)}(t) = -\int_0^t dt_1 Tr_b[L(t)L(t_1)\rho_b] \quad , \tag{2.5.32}$$

$$J^{(1)}(t) = Tr_b L(t)[\rho(0) - \rho_b\sigma(t)] \quad , \tag{2.5.33}$$

$$J^{(2)}(t) = \int_0^t dt_1 \{Tr_b[L(t)L(t_1)(\rho(0) - \rho_b\sigma(0))] -$$

$$-\bar{L}(t)Tr_b L(t_1)(\rho(0) - \rho_b\sigma(0))]\} \quad . \tag{2.5.34}$$

Transforming Eq. (2.5.31) out of interaction representation to the Schrödinger representation and assuming that asymptotic condition (2.4.11) is satisfied we get the Redfield-type equation

$$\dot{\sigma} = -i\hbar^{-1}[E + \bar{V},\sigma] - \Re\sigma \quad . \tag{2.5.35}$$

The inhomogeneous terms $J^{(1)}$ and $J^{(2)}$ decay at times (2.4.11). Equation (2.5.33) differs from Eq. (2.4.8) only by change $\bar{V}^{ef} \to \bar{V}$. Matrix elements of operator $\Re$ (transformed operator R) are determined by Eqs. (2.4.12) and (2.4.2). It follows from the above that the Markovian approximation is achieved provided the weak coupling and asymptotic conditions (2.4.11) and (2.4.7) are satisfied.

To check the validity of Eq. (2.5.35) we have to consider the next order approximation. This is done in Appendix B to this chapter. A general conclusion following from Appendix B is that the third order correction to the decay supermatrix R has the contributions with asymptotic behavior

$$R^{(3)} \propto \lambda^3 t \quad , \tag{2.5.36}$$

where λ is the parameter characterizing the interaction energy V.

Eqs. (2.5.31) and (2.5.35) are valid up to the second order of the parameter λ. The rationale of their validity is that the next order corrections to the supermatrices R and J should have a higher than the second power of λ. This consideration may be true if in the higher approximation ($\propto \lambda^n$, $n > 2$) the combination

$$\lambda^2 t \tag{2.5.37}$$

does not appear; it is not small when $\lambda \to 0$ and $t \to \infty$.

The conclusion is that the validity of Eqs. (2.5.31) and (2.5.35) is not proven, since the next order correction to R, (2.5.36), contains the combination $\lambda^2 t$. The ratio of $R^{(3)}$ to $R^{(2)}$ is not small

$$\frac{R^{(3)}}{R^{(2)}} \sim \frac{\lambda^3 t}{\lambda^2} = \lambda^2 t \cdot \frac{1}{\lambda} \quad . \tag{2.5.38}$$

As a matter of fact, the explanation of the above conclusion is contained in the analysis of the preceding section. It has been shown that equations of the type (2.5.31) and (2.5.35) can be obtained if the parameter r, (2.3.13), is small. For the Hamiltonian (2.5.28), which is the basis of the derivations of the present section, the parameter r is not small. It tends to infinity, since the interaction energy $G \to 0$. In this case the diagonal part of matrix η, (2.5.29), which is the correction to the factorized density matrix $\rho_b \sigma(t)$, (2.3.23), is not small when $t \to \infty$.

Thus, one should be cautious to jump to the conclusion that the next order corrections are small because they are proportional to λ^n, $n > 2$. One should check that the combinations of the $\lambda^2 t$ type do not appear. On the other hand the above conclusions about divergencies (when $G = 0$) in the higher order corrections are based on the diagonality of the dissipative system energy, (2.2.2). In the general case, when F is not diagonal, it implicitly includes the G-type terms. In this case the higher order corrections may be small depending on the magnitude of the off-diagonal terms of F.

2.6 Semigroup theory of irreversible processes

Quantum theory of irreversible processes, which has been presented in the previous section, is based on the von Neumann equation (2.1.1) of the closed system with Hamiltonian (2.1.2). This Hamiltonian describes the system of interest, or dynamic system, whose Hamiltonian is E, interacting with the dissipative system – thermal bath – with Hamiltonian F + G. Using this approach we succeeded to derive the Markovian equation (2.4.14). Equation (2.4.14) without the term $\overline{V}^{ef}$ is usually called the Redfield equation. This equation has certain limitations.

(1) It has been derived in the weak coupling approximation, using only terms up to the second order of the interacting energy V.

(2) Equation (2.4.14) is valid in the factorization approximation (2.3.17).

(3) Equation (2.4.14) has been derived in the Markovian approximation.

There is another approach to the irreversible processes which have been developed during the last decades [39-42]. This approach is based on a number of properties of time development of the density matrix σ of the dynamic system, such as the Markovian property, and positive definiteness of the density matrix σ. Employment of these properties leads to the certain form of the equation for the density matrix σ. Interestingly enough this equation is quite general and does not depend explicitly on the concrete form of the Hamiltonian of the total system (interacting dynamic and dissipative system).

We start from the description of the time behavior of the density matrix σ, assuming that the whole system (dynamic and dissipative) has Hamiltonian H. We do not assume any restrictions on this Hamiltonian. We assume that we are able to prepare at time $t_0 = 0$ the initial state of the total system as an uncorrelated product state

$$\rho(0) = \sigma(0)P_b \ , \tag{2.6.1}$$

where P_b is the fixed reference state of the dissipative system (bath). By U we denote the unitary operator $\exp[-\frac{i}{\hbar}Ht]$ representing the evolution governed by the Hamiltonian H of the total system from $t_0 = 0$ to a certain $t > 0$. Then the transformation describing a state change of the dynamic subsystem may be written as

$$\sigma(t) = \Lambda\sigma(0) = tr_b(U\sigma(0)\cdot P_b U^+) \ , \tag{2.6.2}$$

where tr_b is a partial trace operation over the quantum numbers α of the dissipative system. Without loss of the generality we assume that density matrix P_b is diagonal with matrix elements $P_{\beta\beta}$. Then we obtain the following matrix representation of (2.6.2)

$$(\Lambda\sigma(0))_{mn} = \sum_{\mu,\nu,k,l} P_{\nu\nu} U_{\mu m;\nu k} \sigma_{kl}(0) U^+_{\nu l;\mu n} \ , \tag{2.6.3}$$

where m, n, k, l are indices of the dynamic system.

Hence the <u>dynamic map</u> Λ may be written in the form

$$\Lambda\sigma(0) = \sum_{\alpha} W_\alpha \sigma(0) W^+_\alpha \ , \tag{2.6.4}$$

where W_α acts in the Hilbert space of the dynamic system and

$$\sum W^+_\alpha W_\alpha = 1 \ . \tag{2.6.5}$$

The correspondence between (2.6.3) and (2.6.4) – (2.6.5) is given by

$$\{\alpha\} \equiv \{(\mu,\nu)\}, \qquad (W_\alpha)_{mn} = U_{\mu m;\nu n}\sqrt{P_{\nu\nu}} \ . \tag{2.6.6}$$

To proceed further and to derive a general form of the superoperator Λ we have to introduce quite an abstract notion of the <u>complete positiveness</u> of Λ. First we introduce a superoperator Λ^+ acting on the operators A of the dynamic system (Λ acts on the density matrices). It is clear that

$$tr\{(\Lambda\sigma(0)A\} = tr(\sigma(0)\Lambda^+ A) \ , \tag{2.6.7}$$

where tr operation is performed over indices of the dynamic system. It is easy to show that operator Λ^+ is given by the formula

$$\Lambda^+ A = \sum_\alpha W_\alpha^+ A W_\alpha \tag{2.6.8}$$

Now we discuss necessary mathematical conditions which should be imposed on Λ^+. Because Λ maps density matrices into density matrices at different times, then Λ^+ must be the linear positive – it transforms positive operators into positive ones. Consider now the N-level system with a trivial Hamiltonian H = 0 placed far away from our dynamic system. Hence, because both systems do not interact (and they are assumed to be statistically independent [43]) then the joint dynamical map in the Heisenberg picture Λ_n^+ must be given by a tensor product $\Lambda^+ \otimes 1$ acting on the joint Hilbert space of the dynamic system and the N-level system. Obviously Λ_n^+ should be positive for all n = 1,2,.... This is a new condition on Λ^+ which is called complete positivity. This condition, which does not have an obvious physical explanation, is much stronger than the usual positivity.

We assume the complete positivy of the supermatrix Λ and the quantum semigroup property of Λ, which is expressed by the following relations

a) Λ_t is a dynamic map; it preserves Hermiticity, trace = 1 and is completely positive definite.

b) $\Lambda_t\Lambda_s = \Lambda_{t+s}$.

c) $tr\{(\Lambda_t\sigma(0))A\}$ is a continuous function of t for A belonging to the dynamic system.

Then it can be proven [39-42] that there exists a linear map L, called a generator of a semigroup, such that

$$\frac{d}{dt}\sigma(t) = L\sigma(t) \quad , \tag{2.6.9}$$

where

$$\sigma(t) = \Lambda_t \sigma(0) \quad , \tag{2.6.10}$$

and

$$\Lambda_t = e^{Lt} \quad .$$

Similarly for the Heisenberg dynamics Λ_t^+ we have

$$\frac{dA(t)}{dt} = L^+ A(t) \quad , \tag{2.6.11}$$

with $A(t) = \Lambda_t^+ A(0)$.

Equations (2.6.9) and (2.6.11) are called quantum Markovian equations in the Schrödinger and the Heisenberg picture respectively.

Now, from the complete positivity of Λ and properties a, b and c, follows a general form of supermatrices L and L^+ [41] and corresponding equations of motion

$$\frac{d\sigma}{dt} = L\sigma = -\frac{i}{\hbar}[H,\sigma] + \frac{1}{2}\sum_{\alpha}\{[V_\alpha\sigma, V_\alpha^+] + [V_\alpha, \sigma V_\alpha^+]\} \ , \qquad (2.6.12)$$

$$\frac{dA}{dt} = L^+ A = \frac{i}{\hbar}[H,A] + \frac{1}{2}\sum_{\alpha}\{V_\alpha^+[A, V_\alpha] + [V_\alpha^+, A]V_\alpha\} \ . \qquad (2.6.13)$$

where V_α are certain operators in the Hilbert space of the dynamic system. They are not necessarily Hermitian. On the other hand H is a Hermitian operator $H^+ = H$, which does not necessarily coincide with the Hamiltonian of the dynamic system.

The semigroup approach to quantum dissipation, pioneered by Lindblad [41] provides the most generally allowed form of the equations of motion (2.6.12) and (2.6.13) which satisfy complete positivity. On the other hand we have obtained, in the weak coupling approximation, Markovian equations (2.4.14) which coincide with Redfield equations [29], provided the term $\bar{V}^{ef}$ is omitted.

Pollard and Friesner [44] have shown that if the system-bath interaction is written as a sum of products of the dynamic and dissipative system operators G_α and $\Im_\alpha$ respectively

$$V = \sum_{\alpha} G_\alpha \Im_\alpha \ , \qquad (2.6.14)$$

then the Redfield equations can be written in the form

$$\frac{d\sigma}{dt} = -\frac{i}{\hbar}[E,\sigma] + \frac{1}{\hbar^2}\sum_{\alpha}\{[G_\alpha^+\sigma, G_\alpha] + [G_\alpha, \sigma G_\alpha^-]\} \ , \qquad (2.6.15)$$

where

$$(G_\alpha^+)_{mn} = \sum_{\beta}(G_\beta)_{mn}(\Theta_{\alpha\beta}^+)_{mn} \ , \qquad (2.6.16)$$

$$(G^-)_{mn} = \sum_{\beta}(G_\beta)_{mn}(\Theta_{\beta\alpha}^-)_{mn} \ , \qquad (2.6.17)$$

and

$$(\Theta_{\alpha\beta}^+)_{mn} = \int_0^\infty d\tau e^{-i\omega_{mn}\tau} \langle \Im_\alpha(\tau)\Im_\beta(0)\rangle_\beta \ , \qquad (2.6.18)$$

$$(\Theta^-_{\alpha\beta})_{mn} = \int_0^\infty d\tau e^{-i\omega_{mn}\tau} < \Im_\alpha(0)\Im_\beta(\tau) >_\beta \quad . \tag{2.6.19}$$

Note the striking similarity [45] between the Lindblad form for the equation of motion (2.6.12) and the Redfield equations in the form (2.6.15). However, a closer examination shows [45] that the two formulations are, in general, mutually exclusive, with the Redfield equations not conforming a completely positive semigroup. It does not mean that the Redfield equations of motion cannot preserve positivity for a certain subset of initial states. However, positivity cannot be satisfied for all initial states. A paper [36] of Suarez et al. provides concrete examples of solutions of the Redfield equations which break the positivity; that is, they lead to negative values for population probabilities.

The source of this discrepancy is in the fact that the Markovian approximation is valid only asymptotically, (2.4.11), while the initial conditions are in the region where Markovian Redfield type equations (2.4.14) are not valid. On the other hand, equation (2.4.8) with time dependent coefficients is valid in the whole region $t \geq t_0$. Therefore, solutions of the Markovian Redfield type equation (2.4.14) may violate positive definiteness. Initial conditions for these solutions are located at moments $t \approx t_0$ where the Markovian equation (2.4.14) is not valid (see also [36]). In the derivation of the Lindblad equations (2.6.12) it is assumed that the Markovian property takes place at all times starting from the initial moment. Such an assumption could not be justified on the physical basis. In the region close to the initial moment characterized by the factorized density matrix (2.6.1) the time dependence is not Markovian.

The semigroup approach does not employ the weakness of interaction between the dynamic and dissipative system. On the other hand it employs property b) of the time development

$$\Lambda_{t+s} = \Lambda_t \Lambda_s \quad . \tag{2.6.20}$$

This condition is connected with the validity of the factorization approximation. Using Eq. (2.6.2), the action of superoperator Λ_{t+s} can be written in the form

$$\Lambda_{t+s}\sigma(0) = \mathrm{tr}_b \left\{ e^{-\frac{i}{\hbar}Ht} \left[e^{-\frac{i}{\hbar}Hs} \sigma(0) P_b e^{\frac{i}{\hbar}Hs} \right] e^{\frac{i}{\hbar}Ht} \right\} \quad . \tag{2.6.21}$$

Now let us assume that a condition similar to (2.3.13) is satisfied. Then the factorization approximation is valid and the square bracket may be presented as in the right hand side of Eq. (2.6.21)

$$e^{-\frac{i}{\hbar}Hs} \sigma(0) P_b e^{\frac{i}{\hbar}Hs} = \sigma(s) P_b = \Lambda_s \sigma(0) P_b \quad . \tag{2.6.22}$$

Substituting this expression into Eq. (2.6.21) we get

$$\Lambda_{t+s}s(0) = tr_b\left\{e^{-\frac{i}{\hbar}Ht}\sigma(s)\cdot P_b e^{\frac{i}{\hbar}Ht}\right\} = \Lambda_t\sigma(s) = \Lambda_t\Lambda_s\sigma(0) \quad . \tag{2.6.23}$$

This equation is equivalent to Eq. (2.6.20). In conclusion, we have to emphasize that Eq. (2.6.12) is not self-evident, though it has been derived using quite general assumptions.

The weak point of the derivation is the assumption that the semigroup property is valid at the whole time interval $t \geq 0$, while Λ_t is determined by Eqs. (2.6.9, 2.6.10) with time-independent L.

2.7 Master equations for dynamic systems

Now we will derive conditions under which the equations for the density matrix of the dynamic system (2.4.14) can be reduced to the conventional master equations. Equation (2.4.14) can be rewritten in the explicit form

$$\dot{\sigma}_{mn} = -i\omega_{mn}\sigma_{mn} - \frac{i}{\hbar}\sum_{l}\left(\overline{V}^{ef}_{ml}\sigma_{ln} - \sigma_{ml}\overline{V}^{ef}_{ln}\right) - \sum_{m'n'}\Re^{m'n'}_{mn}\sigma_{m'n'} \quad . \tag{2.7.1}$$

We will transform Eq. (2.7.1) to the interaction representation

$$\sigma_{mn} = \overline{\sigma}_{mn}e^{-i\omega_{mn}t} \quad , \tag{2.7.2}$$

and obtain the equation for $\overline{\sigma}_{mn}$

$$\dot{\overline{\sigma}}_{mn} = -i\hbar^{-1}\left(\overline{V}^{ef}_{mk}\overline{\sigma}_{kn}e^{i\omega_{mk}t} - \overline{\sigma}_{mk}\overline{V}^{ef}_{kn}e^{i\omega_{kn}t}\right)$$

$$-\sum_{m'n'}\Re^{m'n'}_{mn}\overline{\sigma}_{m'n'}e^{i(\omega_{mn}-\omega_{m'n'})t} \quad . \tag{2.7.3}$$

Now we will make the following assumption about the energy spectrum of the system.

(a) All the levels of the dynamic system are non-degenerate:

$$\omega_{mn} \neq 0; \qquad \text{if } m \neq n \ ,$$

(b)
$$|\omega_{mn}| >> \frac{|\overline{V}^{ef}_{mn}|}{\hbar} , \tag{2.7.4}$$

where T_{dyn} is the characteristic time of the relaxation of the dynamic subsystem. It has the order of magnitude

$$T_{dyn} \sim \left|\Re_{mn}^{m'n'}\right|^{-1} . \tag{2.7.5}$$

Equation (2.7.3) contains terms which do not depend on time explicitly (e.g. a term with $\omega_{mn} = \omega_{m'n'}$, in the right hand side of (2.7.3)), the so-called secular terms, and rapidly varying terms proportional to $e^{i\omega_{mn}t}$. It is easy to show that the contribution of these rapidly varying terms is much less than those which are independent of time.

In this approximation, which is sometimes called a secular approximation, the equations for the diagonal elements σ_{nn} (probabilities to find the system in state $|n>$) obtain the form

$$\dot{\sigma}_{nn} = -\sum_k \Re_{nn}^{kk} \sigma_{kk} . \tag{2.7.6}$$

Using Eqs. (2.4.15) - (2.4.17) we get

$$\Re_{nn}^{nn} = \hbar^{-2} \sum_{k(\neq n)} \int_{-\infty}^{\infty} \left\langle \tilde{V}_{nk}(\tau)\tilde{V}_{kn}(0)\right\rangle e^{-i\omega_{kn}\tau} d\tau = \sum_{k(\neq n)} w_{nk} , \tag{2.7.7}$$

$$\underset{(k\neq n)}{\Re_{nn}^{kk}} = -\hbar^{-2} \int_{-\infty}^{\infty} \left\langle \tilde{V}_{kn}(\tau)\tilde{V}_{nk}(0)\right\rangle e^{-i\omega_{nk}\tau} d\tau = -w_{kn} . \tag{2.7.8}$$

Here

$$w_{nk} = \frac{2\pi}{\hbar} \sum_{\substack{\alpha,\alpha' \\ (\alpha\neq\alpha')}} P_\alpha \left|\tilde{V}_{n\alpha;k\alpha'}\right|^2 \delta(F_\alpha - F_{\alpha'} - E_k + E_n)$$
$$= \hbar^{-2} \int_{-\infty}^{\infty} \left\langle \tilde{V}_{nk}(\tau)\tilde{V}_{kn}(0)\right\rangle e^{-i\omega_{kn}\tau} d\tau . \tag{2.7.9}$$

From the above equations we obtain the master equation for the dynamic system

$$\dot{P}_n = -\sum_k \left(w_{nk} P_n - w_{kn} P_k \right) , \tag{2.7.10}$$

where P_n is the probability to find the system in the state n. This probability coincides with the diagonal element of the dynamic system density matrix

$$P_n = \sigma_{nn} . \tag{2.7.11}$$

If the dissipative system is in the state of thermal equilibrium then

$$P_\alpha = e^{-F_\alpha/k_B T} / \sum_\beta e^{-F_\beta/k_B T} . \tag{2.7.12}$$

From this expression and Eq. (2.7.9) we obtain the relation between the transition probabilities $n \to k$ and $k \to n$

$$w_{nk} = w_{kn} \exp[-(E_n - E_k)/k_B T] \ . \qquad (2.7.13)$$

This relation ensures the Boltzmann distribution

$$P_n = e^{-E_n/k_B T} / \sum_k e^{-E_k/k_B T} \ , \qquad (2.7.14)$$

to be the asymptotic solution ($t \to \infty$) of the master equation (2.7.10).

Simple equations for the off-diagonal elements σ_{mn} of the dynamic system density matrix may be derived, if one makes an additional assumption about the non-degeneracy of frequencies ω_{mn}:

$$\omega_{mn} \neq \omega_{m'n'}, \quad \text{if} \quad (m,n) \neq (m',n')$$

and that

$$|\omega_{mn} - \omega_{m'n'}| >> |\bar{V}^{cf}|/\hbar, \ \ T_{dyn}^{-1}; \ (m,n) \neq (m',n') \ . \qquad (2.7.15)$$

Then using the same argument as in the derivation of the master equation (2.7.10), we take into account only those terms in Eq. (2.7.3) which do not have explicit dependence on time. As a result we obtain

$$\dot{\sigma}_{mn} + i\hbar^{-1}[E + \bar{V}^{cf} + \Gamma, \sigma]_{mn} + \frac{1}{\tau_{mn}} \sigma_{mn} = 0 \ , \ \ (m \neq n) \ . \qquad (2.7.16)$$

Here

$$\tau_{mn}^{-1} = \frac{1}{2} \sum_k (w_{mk} + w_{nk}) + \frac{\pi}{\hbar} \sum_{\alpha,\alpha'} P_\alpha |V_{m\alpha';m\alpha} - V_{n\alpha';n\alpha}|^2 \, \delta(F_\alpha - F_{\alpha'}) \ , \qquad (2.7.17)$$

$$\Gamma_{mk} = \hbar^{-1} \sum_{\substack{l,\alpha,\alpha' \\ (\alpha \neq \alpha')}} P_\alpha \frac{V_{m\alpha;l\alpha'} V_{l\alpha';m\alpha}}{\omega_{lm} + \omega_{\alpha\alpha'}} \delta_{mk} \ . \qquad (2.7.18)$$

Since $\sigma_{mn}(t)$ describes harmonic oscillations between two levels m and n, the quantities τ_{mn} describe the relaxation time of these oscillations. The first term in the right hand side of Eq. (2.7.17) is connected with the relaxation of populations of levels m and n (w_{mk} and w_{nk}), while the second term describes so-called pure dephasing. The pure dephasing occurs even when there is no population relaxation from levels m and n.

We shall now consider two examples: dissipation of the harmonic oscillator and of the two level system. The harmonic oscillator has equidistant energy levels and therefore it satisfies conditions (2.7.4) if the difference between the levels of the oscillator $\hbar\omega_0$ is large enough. On the other hand its eigenfrequencies are degenerate $\omega_{n+1,n} = \omega_{n+2,n+1} = \omega_0$ and therefore condition (2.7.15) is not satisfied. The two level system satisfies both conditions (2.7.4) and (2.7.5) provided the energy difference between the two levels is large enough.

Let us first consider the harmonic oscillator. Dissipation of the harmonic oscillator has been considered in Refs. 46-48 and in the most extensive study of Dekker [49]. We will present here the derivation of the "master" equation for the harmonic oscillator, using the Redfield equation in the form (2.6.15). We assume that the interaction energy of the harmonic oscillator with the dissipative system can be written in the form

$$V = \sum_{\alpha} q_{\alpha} F_{\alpha} \ , \tag{2.7.19}$$

where F_α are operators of the dissipative system while the operator q is the operator of the harmonic oscillator

$$q_{\alpha} = q = a + a^{+} \ , \tag{2.7.20}$$

and a and a^+ are annihilation and creation operators respectively

$$a\psi_n = \sqrt{n}\psi_{n-1}; \qquad a^{+}\psi_n = \sqrt{n+1}\psi_{n+1} \ , \tag{2.7.21}$$

and ψ_n are eigenfunctions of the harmonic oscillator.

Substituting (2.7.19) and (2.7.20) into Eqs. (2.6.15-2.6.19) (and taking into account the $\bar{V}^{ef}$ term) we get

$$\frac{d\sigma}{dt} = \frac{1}{\hbar}[E + \bar{V}^{ef}, \sigma] + \frac{1}{\hbar^2}\{[(qx\theta^{+})\sigma, q] + [q, \sigma(qx\theta^{-})]\} \ . \tag{2.7.22}$$

Here, according to (2.6.18 – 2.6.19)

$$(qx\theta^{+})_{mn} = q_{mn}\theta^{+}_{mn}; \quad (qx\theta^{-})_{mn} = q_{mn}\theta^{-}_{mn} \ , \tag{2.7.23}$$

and

$$\theta^{+}_{mn} = \sum_{\alpha,\beta}\int_0^{\infty} e^{-i\omega_{mn}\tau} \langle F_{\alpha}(\tau)F_{\beta}(0)\rangle d\tau \ , \tag{2.7.24}$$

$$\theta^{-}_{mn} = \sum_{\alpha,\beta}\int_0^{\infty} e^{-i\omega_{mn}\tau} \langle F_{\alpha}(0)F_{\beta}(\tau)\rangle d\tau \ . \tag{2.7.25}$$

Taking into account Eqs. (2.7.20 – 2.7.21) and that the spectrum of the eigenfrequencies of the harmonic oscillator has the form

$$\omega_{mn} = \pm\omega_0 \ , \tag{2.7.26}$$

we obtain

$$(qx\theta^+) = a\theta^+_- + a^+\theta^+_+ ; \qquad (qx\theta^-) = a\theta^-_- + a^+\theta^-_+ ; \tag{2.7.27}$$

where

$$\theta^+_\pm = \sum_{\alpha,\beta} \int_0^\infty d\tau e^{\mp i\omega_0\tau} <F_\alpha(\tau)F_\beta(0)> \ ;$$

$$\theta^-_\pm = \sum_{\alpha,\beta} \int_0^\infty d\tau e^{\mp i\omega_0\tau} < F_\alpha(0)F_\beta(\tau)> . \tag{2.7.28}$$

Thus the equation for the density matrix of the harmonic oscillator interacting with the dissipative system has the form

$$\frac{d\sigma}{dt} = -\frac{i}{\hbar}[E + \overline{V}^{ef}, \sigma] + \frac{1}{\hbar^2}\{[(a\theta^+_- + a^+\theta^+_+)\sigma, q] + [q, \sigma(a\theta^-_- + a^+\theta^-_+)]\} \ . \tag{2.7.29}$$

In the above-mentioned secular approximation the terms proportional to a^+a^+ and aa are neglected and Eq. (2.7.29) obtains the form

$$\frac{d\sigma}{dt} = -\frac{i}{\hbar}[E + \overline{V}^{ef}, \sigma] + \frac{1}{\hbar^2}\{[a\theta^+_-\sigma, a^+] + [a^+\theta^+_+\sigma, a]$$

$$+ [a, \sigma a^+\theta^-_+] + [a^+, \sigma a\theta^-_-]\} \ . \tag{2.7.30}$$

The above approximation in which the so-called counter-rotating terms a^+a^+ and aa are neglected is also known as the rotating wave approximation (RWA). A more detailed discussion of RWA is performed in Section 5.7.

It is easy to see that RWA equation (2.7.30) connects diagonal terms $\dot{\sigma}_{nn}$ with diagonal terms σ_{nn}, $\sigma_{n\pm1;n\pm1}$ only. It means that Eq. (2.7.30) is a master equation (2.7.10) as far as the diagonal part of the left-hand side of Eq. (2.7.30) is considered. On the other hand the off-diagonal matrix elements in Eq. (2.7.30) or (2.7.29) do not obey those of type of equation (2.7.16). The harmonic oscillator eigenfrequencies are degenerate and do not satisfy condition (2.7.15).

We now consider equations of motion of the density matrix of the two-level system. In this case, as in the case of the harmonic oscillator (2.7.29), we can obtain equations without assuming the rotating wave approximation [17,18]. Using Eqs. (2.4.14 – 2.4.17) we obtain the following equations of motion

$$\dot{\sigma}_{11} = \dot{\sigma}_{22} = -i\hbar^{-1}(\overline{V}_{12}^{cf}\sigma_{21} - \sigma_{12}\overline{V}_{21}^{cf}) - (w_{12}\sigma_{11} - w_{21}\sigma_{22}) , \qquad (2.7.31)$$

$$\dot{\sigma}_{12} = \dot{\sigma}_{21}^{*} - -i\omega_{12}\sigma_{12} - i\hbar^{-1}\overline{V}_{12}^{ef}(\sigma_{22} - \sigma_{11}) - R\sigma_{12} + R^{*}\sigma_{21} . \qquad (2.7.32)$$

Here $\overline{V}_{12}^{ef}$ is determined by Eq. (2.4.9), $\overline{V}_{11}^{ef} = \overline{V}_{22}^{ef} = 0$, w_{12}, w_{21} are transition probabilities per unit time (2.7.9) and quantities R, R* by Eqs. (2.4.14 – 2.4.17)

$$R_{12}^{12} = \frac{2}{\hbar^2}\int_0^{\infty} d\tau e^{-i\omega_{21}\tau}\,\mathrm{Re}\{<\tilde{V}_{12}(\tau)\tilde{V}_{21}(0)>\} \equiv R = -(R_{12}^{21})^{*} = -R^{*} . \qquad (2.7.33)$$

It is easy to see that terms with $\overline{V}_{12}^{ef}$ are not secular and can be neglected, provided condition (2.7.4) is satisfied. The term $R^{*}\sigma_{21}$ is explicitly counterrotating and may also be neglected if condition (2.7.4) is satisfied. Thus in the RWA Eqs. (2.7.31 – 2.7.32) coincide with Eqs. (2.7.10) and (2.7.16) with

$$\tau_{12}^{-1} = \mathrm{Re}(R) . \qquad (2.7.34)$$

On the other hand if $|\omega_{12}|$ is small enough, and conditions (2.7.4) and (2.7.15) (they coincide for the two-level system) are not satisfied, then terms with $\overline{V}^{ef}$ and R^{*} are essential.

Appendix A. The factorization approximation.

The term $\eta_{mn\alpha}(t)$ in Eq. (2.3.15) represents a statistical correlation between the dynamic and dissipative subsystems. We will show that if this correlation is small [in comparison with the first term in the right hand side of Eq. (2.3.15)], then we can get the equation for the density matrix $\sigma(t)$ only. It is clear from physical considerations that very fast relaxation of the dissipative system would destroy the correlation $\eta_{mn\alpha}$ created by the interaction V between the dynamic and the dissipative subsystems. Substituting the solution $\rho_{m\alpha;n\alpha}(t)$ in the form (2.3.15) into equation (2.3.11) and performing the transformation

$$\sigma_{mn}, \eta_{mn\alpha} \to \bar{\sigma}_{mn} e^{-i\omega_{mn}t}, \quad \bar{\eta}_{mn\alpha} e^{-i\omega_{mn}t}, \tag{2.A1}$$

we obtain the approximate equations for the correlation amplitudes $\bar{\eta}_{mn\alpha}$

$$\dot{\bar{\eta}}_{mn\alpha} + \sum_{\beta} w_{\alpha\beta}(\bar{\eta}_{mn\alpha} - \bar{\eta}_{mn\beta}) = -i \sum_{l,k,\alpha',\beta'} \{D L \exp[-it(t-t_0)(1-D)L]\}_{m\alpha;n\alpha}^{l\alpha';k\beta'}$$

$$x\rho_{l\alpha';k\beta'}(t_0)(1-\delta_{\alpha'\beta'}) - \dot{\bar{\sigma}}_{mn} P_\alpha - \frac{i}{\hbar} P_\alpha \sum_k \left(V_{m\alpha;k\alpha} \bar{\sigma}_{kn} e^{i\omega_{mk}t} - \bar{\sigma}_{mk} V_{k\alpha;n\alpha} e^{i\omega_{kn}t} \right)$$

$$-\sum_{lk\beta} \int_0^{t-t_0} d\tau \left[\bar{K}^{VV}(\tau) + \bar{K}^{\sigma V}(\tau) \right]_{m\alpha;n\alpha}^{l\beta;k\beta} \bar{\sigma}_{lk}(t-\tau) e^{i\omega_{lk}\tau} e^{i(\omega_{mn}-\omega_{lk})t} P_\beta \,. \tag{2.A2}$$

In this equation we have neglected all the terms proportional to $\bar{\eta}_{lk\alpha}$ [apart from the second sum in the l.h.s. of Eq. (2.A2)]. They are small in comparison with the second term in the left-hand side of Eq. (2.A2). These terms are proportional to powers of small parameter r, (2.3.14), while the term $\sum_{\beta} w_{\alpha\beta}(\bar{\eta}_{mn\alpha} - \bar{\eta}_{nm\beta})$ does not depend on this parameter.

Employing the analysis of Reference 38, the left hand side of Eq. (2.A2) may be presented as

$$\tilde{\eta}_{mn} + \tilde{A}\tilde{\eta}_{mn} \quad \text{with} \quad \tilde{A}_{\alpha\beta} = \delta_{\alpha\beta} \sum_{\delta} w_{\delta\alpha} - (1-\delta_{\alpha\beta}) w_{\alpha\beta} \;, \tag{2.A3}$$

where $\tilde{\eta}_{mn}$ is the vector and $\tilde{A}$ is the operator in Hilbert space (characterized by quantum numbers α, and with fixed indices m,n). The operator $\tilde{A}$ has non-negative eigenvalues. The first (smallest) non-zero eigenvalue of operator $\tilde{A}$ is the relaxation rate γ. Other eigenvalues $\lambda_n > \gamma$ lead to the fast transient parts of $\bar{\eta}_{mn\alpha}$. In the relaxation approximation, taking into account only the eigenvalue $\lambda_1 = \gamma$, one can present the solution of equation (2.A2) in the form (assuming $\bar{\eta}_{mn\alpha}(t_0) \neq 0$))

$$\overline{\eta}_{mn\alpha}(t) = e^{-\gamma t}\int_{t_0}^{t} R(\tau)e^{\gamma\tau}d\tau + \overline{\eta}_{mn\alpha}(t_0)e^{-\gamma(t-t_0)} , \qquad (2.A4)$$

where R(t) is the right hand side part of equation (2.A2).

First, let us estimate the contribution of the first term R(t) in the right hand side of equation (2.A2). We assume that $R_1(t)$ decays exponentially, γ_c being the characteristic decay rate

$$R_1(t) = R_1(t_0)e^{-\gamma_c(t-t_0)} . \qquad (2.A5)$$

We again assume, as it has been done in the previous section [cf (2.2.31)] that

$$\gamma_c >> \gamma = T_{diss}^{-1} . \qquad (2.A6)$$

The contribution of other terms may be represented by the second term in the right hand side of Eq. (2.A2) - $\dot{\overline{\sigma}}_{mn}P$. This term has the same order as the sum of the next terms. (From the below consideration we will see that the sum over α of all the terms in the right hand side of Eq. (2.A2), apart from the first term, is approximately equal to zero.)

Having in mind the conditions (2.3.13) and (2.3.14) we can present the estimate of $\overline{\eta}_{mn}(t)$, (2.A4), as

$$\overline{\eta}_{mn\alpha}(t) \approx \overline{\eta}_{mn\alpha}(t_0)e^{-\gamma(t-t_0)} - R_1(t_0)\frac{e^{-(\gamma_c-\gamma)(t-t_0)}-1}{\gamma_c-\gamma}e^{-\gamma(t-t_0)}$$

$$-\dot{\overline{\sigma}}_{mn}P_\alpha\frac{1-e^{-\gamma(t-t_0)}}{\gamma} . \qquad (2.A7)$$

Starting from small interval

$$t - t_0 \gtrsim \frac{1}{\gamma} = T_{diss} , \qquad (2.A8)$$

after the initial time t_0, the correlation factor $\overline{\eta}_{mn\alpha}$ becomes small in comparison with $\overline{\sigma}_{mn} \cdot P_\alpha$

$$\overline{\eta}_{mn\alpha} \sim \overline{\sigma}_{mn}P_\alpha\frac{T_{diss}}{T_{dyn}} , \qquad (2.A9)$$

provided condition (2.3.13) is satisfied. Taking initial time t_0 such that

$$\bar{\eta}_{mn\alpha}(t_0) = 0; \qquad \rho_2(t_0) = 0 \;\; [R_1(t_0) = 0] \; . \tag{2.A10}$$

one obtains that the factorization approximation

$$\rho_{m\alpha;n\alpha}(t) \approx \sigma_{mn}(t) P_\alpha \;\; , \tag{2.A11}$$

is valid at any time $t - t_0 > 0$. In the general case, the factorization approximation is valid at the time intervals (2.A8), provided the condition (2.3.13) is satisfied. Of course it does not mean that the Markovian equations (2.4.14) with constant coefficients are valid starting from t_0. In the region $t\text{-}t_0 \sim \tau_c$ equations (2.4.8) with time-dependent coefficients take place.

Appendix B. The third order correction in time convolutionless equations.

A third order correction to the time dependent R-function in Eq. (2.5.30) has the form

$$R^{(3)}(t) = i\rho_b^{-1} P L(t) \eta^{(2)}(t) \quad . \tag{2.B1}$$

Here the η-function is defined as

$$\eta(t) = Q\rho(t); \qquad \eta(0) = 0 \quad . \tag{2.B2}$$

This function is similar to function $\eta_{mn\alpha}(t)$, (2.5.29), but it has also off-diagonal elements in Greek indices α,β

$$\eta_{m\alpha;n\beta}(t) = \rho_{m\alpha;n\beta}(t) - \sigma_{mn}(t) P_\alpha \delta_{\alpha\beta} \quad . \tag{2.B3}$$

Function $\eta_{m\alpha;n\alpha}(t)$ coincides with $\eta_{mn\alpha}(t)$.

According to Eqs. (2.5.15) and (2.5.23), the second order correction $\eta^{(2)}$ has the form (with $\eta(0) = 0$)

$$\eta^{(2)}(t) = \sum\nolimits^{(2)}(t) P\rho(t) \quad . \tag{2.B4}$$

Performing some identical transformations, we can rewrite $\eta^{(2)}$ in the form

$$\eta^{(2)}(t) = \int_0^t dt_1 \left[\hat{R}(t_1) - \rho_b R(t_1)\right] \sigma(t)$$

$$+ \int_0^t dt_1 \int_0^{t_1} dt_2 \rho_b \left[\bar{L}(t_1)\bar{L}(t_2) + \bar{L}(t_2)\bar{L}(t_1)\right] \sigma(t)$$

$$- \int_0^t dt_1 \int_0^{t_1} dt_2 \left[L(t_1)\rho_b \bar{L}(t_2) + L(t_2)\rho_b \bar{L}(t_1)\right] \sigma(t) \quad . \tag{2.B5}$$

Here

$$\bar{L}(t)\sigma = \frac{1}{\hbar}\left[\bar{V}(t), \sigma\right]; \qquad \bar{L}(t) = Tr_b L(t)\rho_b \ ; \tag{2.B6}$$

$$\hat{R}(t) = \int_0^t dt_1 L(t) L(t_1) \rho_b;$$

$$R(t) = R^{(2)}(t) = \int_0^t dt_1 Tr_b L(t)L(t_1)\rho_b = Tr_b \hat{R}(t) \quad . \tag{2.B7}$$

It is easy to check that condition (2.3.15) is satisfied and

$$Tr_b \eta^{(2)}(t) = 0 \quad . \tag{2.B8}$$

The trace, (Tr_b), of the first integral in the right hand side of Eq. (2.B5) is zero due to Eqs. (2.B7). The trace of the sum of the second and third integrals in the right hand side of Eq. (2.B5) is zero due to Eqs. (2.B6). The first integral in the right hand side of Eq. (2.B5) and the rest terms in the right hand side of the equation have different time dependencies. The first integral contains the correlation of type

$$<V(t)V(t_1)> \quad . \tag{2.B9}$$

while the rest of the terms in the right hand side of Eq. (2.B5) has the time dependencies connected with the mean values

$$< V(t) >= \bar{V}(t) \quad . \tag{2.B10}$$

We will show now that $\eta^{(2)}(t)$ contains divergent terms of the type

$$V^2 t \sim \lambda^2 t \quad . \tag{2.B11}$$

In this case the second order approximation, which is the basis of the Redfield type equation (2.5.35), is not valid. The next order approximation $R^{(3)}$ is not small in the case (2.B11)

The operator R(t), in the interaction representation, may be presented as

$$R^{lk}_{mn}(t) = \Re^{lk}_{mn}(t) e^{i(\omega_{mn} - \omega_{lk})t} \quad , \tag{2.B12}$$

where operator $\Re^{lk}_{mn}(t)$

$$\Re^{lk}_{mn}(t) = \int_0^t (K^{VV}(\tau))^{lk}_{mn} e^{i\omega_{lk}\tau} d\tau \quad ; \tag{2.B13}$$

and $\bar{K}^{VV}(\tau)$ is determined by Eq. (2.4.2). When $t \to \infty$, $\Re^{lk}_{mn}(t)$ tends to constant supermatrix (2.4.12).

The integral

$$\int_0^t dt_1 R(t_1) = \int_0^t \Re^{lk}_{mn}(t_1) e^{i(\omega_{mn} - \omega_{lk})t} \quad , \tag{2.B14}$$

which appears in the first term of the right hand side of Eq. (2.B5) is divergent when

$$\omega_{mn} - \omega_{lk} = 0 \qquad \text{and} \qquad t \to \infty . \tag{2.B15}$$

The value of the integral (2.B14) may then be estimated as

$$\int_0^{n\tau_c} \left[\Re_{mn}^{lk}(t) - \Re_{mn}^{lk} \right] d\tau + \Re_{mn}^{lk} t \to \Re_{mn}^{lk} t , \tag{2.B16}$$

where $\tau = n\tau_c$, $n >> 1$, and τ_c is the correlation time characterizing the correlation function $\bar{K}^{VV}(\tau)$, while the constant supermatrix $\Re_{mn}^{lk}$ is determined by Eq. (2.4.12). Matrix elements of operator $\hat{R}(t_1)$ in the first integral of Eq. (2.B5) cannot cancel the divergency (2.B16). Only the trace (tr_b) annul the first term in the right hand side of Eq. (2.B5).

The second integral of Eq. (2.B5) is the sum of the terms having the form

$$\rho_b \int_0^t dt_1 \int_0^{t_1} dt_2 \sum_{k,l} \bar{V}_{mk} \bar{V}_{kl} \sigma_{ln}(t) e^{i\omega_{mk} t_1} e^{i\omega_{kl} t_2}$$

$$= \rho_b \sum_{kl} \frac{\bar{V}_{mk} \bar{V}_{kl}}{i\omega_{kl}} \left[\frac{e^{i\omega_{ml} t} - 1}{i\omega_{ml}} - \frac{e^{i\omega_{mk} t} - 1}{i\omega_{mk}} \right] \sigma_{ln}(t) . \tag{2.B17}$$

This term is diverging ($t \to \infty$) if one of the conditions (or both of them) is satisfied

$$\omega_{ml} = 0 \; ; \; \omega_{mk} = 0 . \tag{2.B18}$$

Divergencies connected with the second integral of Eq. (2.B5) cannot be compensated by the third integral. Only the trace operation (tr_b) cancels the sum of the second and third integrals of Eq. (2.B5).

CHAPTER III

INTERACTION WITH PHONONS AND MOLECULAR VIBRATIONS

In this chapter we present a theoretical application of the general formalism developed in the preceding chapter. The main subject of this chapter is the interaction of the two-state system with a harmonic phonon bath. The two-state system is a dynamic system (according to the terminology of the preceding chapter), while the phonon bath is a dissipative system. The role of the phonon bath may play the vibrations in condensed media or huge molecules. These vibrations have an infinite (in the limit) number of degrees of freedom and a continuum spectrum of energies – the continuum of frequencies of phonons. The model of the two-state system interacting with a continuum of the vibrations, may represent various physical situations, such as radiationless transitions in huge molecules, electron and energy transfer, tunneling, and other processes in condensed media, such as small polaron motion, group transfer in biological systems, etc. Equations of motion of two-level systems in condensed media, as well as the calculation of the rate coefficients, are the subject of the chapter. The mutual influence of a large number of two-state systems and the vibrations of the phonon bath are also analyzed in this chapter. This mutual influence may cause the instabilities in the phonon bath. The latter phenomenon, although it has few experimental implementations, is of theoretical importance.

3.1 Description of time-dependent electron-nuclear system in the Born-Oppenheimer approximation

Chemical compounds, molecules – small and large (like proteins) as well as solids – all of them may be presented as a set of interacting electrons and nuclei. In the non-relativistic approximation the Hamiltonian of the system of electrons and nuclei can be written as

$$\mathcal{H} = \mathrm{T} + \mathrm{H}(q,\xi)\,, \tag{3.1.1}$$

where T is the kinetic energy of the nuclei

$$\mathrm{T} = \sum_k \frac{p_k^2}{2M_k} = -\frac{\hbar^2}{2}\sum_k \frac{1}{M_k}\frac{\partial^2}{\partial q_k^2}\ . \tag{3.1.2}$$

Here p_k and q_k are momenta and coordinates of the nuclei, M_k are their masses, and

$$\mathrm{H}(q,\xi) = \sum_j \frac{\pi_j^2}{2m} + \mathrm{U}(\xi,q)\ , \tag{3.1.3}$$

is the electronic Hamiltonian which depends on nuclear coordinates $q = \{q_1, q_2, \ldots, q_n, \ldots, q_N\}$; π_j is the momentum of the j-th electron, m is mass of the electron, $\xi = \{\xi_1, \xi_2, \ldots, \xi_j, \ldots\}$ are the coordinates of the electrons and $U(\xi,q)$ represents the Coulomb interaction between the electrons and nuclei.

Due to the fact that the mass m of the electrons is much smaller than masses of the nuclei M_k it is possible to employ the so-called adiabatic approximation or the Born-Oppenheimer approximation [50]. In this method we may approximately calculate the electronic eigenfunctions, not taking into account the kinetic energy of nuclei, the latter being considered as a small perturbation. Describing the electronic motion, we assume that it is possible to neglect the motion of nuclei and that the electronic eigenfunctions depend on nuclear coordinates as on parameters. Thus in the zero approximation

$$H_0(q,\xi)\chi_\alpha(q,\xi) = E_\alpha(q)\chi_\alpha(q,\xi) \quad . \tag{3.1.4}$$

(The index "0" shows that H_0 does not necessarily coincide with H. However, it will be assumed that the difference $H-H_0$ is small.) In the adiabatic approximation it is assumed also that the nuclear wave functions satisfy the equation

$$\left(H_{\alpha\alpha}(q) + T_{\alpha\alpha}(q) + T\right)\varphi_{\alpha m}(q) = E_{\alpha m}(q)\varphi_{\alpha m}(q) \quad , \tag{3.1.5}$$

where $\mathcal{H}_{\alpha\alpha} = H_{\alpha\alpha} + T_{\alpha\alpha}$ is the Hamiltonian (3.1.1) of the system averaged over the electronic motions:

$$H_{\alpha\alpha}(q) = \int d\xi \chi_\alpha^* H(\xi,q)\chi_\alpha \quad , \tag{3.1.6}$$

$$T_{\alpha\alpha}(q) = -\frac{\hbar^2}{2}\sum_k \frac{1}{M_k}\int d\xi \chi_\alpha^* \frac{\partial^2 \chi_\alpha}{\partial q_k^2} \quad . \tag{3.1.7}$$

Thus in the adiabatic approximation the nuclear motion is described by the Hamiltonian depending on nuclear coordinates and the role of the potential energy is played by the total Hamiltonian of the system averaged over the electronic motion. The effective potential energy of the nuclei is

$$U_\alpha(q) = H_{\alpha\alpha}(q) + T_{\alpha\alpha}(q) \quad , \tag{3.1.8}$$

and it depends on the electronic state. In different electronic states $|\alpha\rangle$ there are different potential energies.

The wavefunction of the total system, nuclei and electrons may be presented as

$$\psi = \sum_{\alpha m} c_{\alpha m}\varphi_{\alpha m}(q)\chi_\alpha(q,\xi)e^{-iE_{\alpha m}t/\hbar} \quad . \tag{3.1.9}$$

In the zero approximation, coefficients $c_{\alpha m}$ do not depend on time. To show the validity of the adiabatic approximation we have to show the slowness of the time dependence of $c_{\alpha m}(t)$. We substitute (3.1.9), with time-dependent $c_{\alpha m}(t)$, into the Schrödinger equation of motion

$$i\hbar\frac{\partial\psi}{\partial t}=H\psi\,. \tag{3.1.10}$$

(The use of the matrix density formalism does not change the results of the derivation.)

After substitution of (3.1.9) into (3.1.10) we multiply both parts of the resulting equation by $\psi_\alpha(q,\xi)\varphi_{\alpha n}(q)$ and integrate over the variables ξ and q. Using the orthonormality of these functions and Eq. (3.1.5) we get the equation for $c_{\alpha n}$

$$i\hbar\dot{c}_{\alpha n}=\sum_{m,\beta}<n\,|\,A_{\alpha\beta}\,|\,m>\exp[i(E_{\alpha n}-E_{\beta m})t/\hbar]\,. \tag{3.1.11}$$

Here

$$A_{\alpha\beta}(q)=H_{\alpha\beta}(1-\delta_{\alpha\beta})+T_{\alpha\beta}(1-\delta_{\alpha\beta})-\hbar^2\sum_i\frac{1}{M_i}\left(\frac{\partial}{\partial q_i}\right)_{\alpha\beta}\frac{\partial}{\partial q_i}\,, \tag{3.1.12}$$

and

$$\left(\frac{\partial}{\partial q_j}\right)_{\alpha\beta}=\int\chi_\alpha^*\frac{\partial\chi_\beta}{\partial q_i}d\xi\,. \tag{3.1.13}$$

In the case when the electronic eigenfunctions may be chosen real (e.g., in the absence of the external magnetic field), the diagonal matrix elements

$$\left(\frac{\partial}{\partial q_j}\right)_{\alpha\alpha}=\frac{1}{2}\frac{\partial}{\partial q_j}\int\chi_\alpha^2d\xi=0\,. \tag{3.1.14}$$

Operator A with matrix elements (3.1.12) plays the role of the effective interaction energy. The adiabatic approximation is justified when matrix elements of A are small. The second and third terms of A, (3.1.12), are small due to the assumption that $M_l >> m$. The smallness of the first term of (3.1.12) depends on the choice of Hamiltonian H_0 in Eq. (3.1.4). In the case when matrix elements of $H-H_0$ are small enough the first term of Eq. (3.1.12) is small too. In the case of infinite masses M_k and $H_{\alpha\beta}(1-\delta_{\alpha\beta})=0$, the coefficients $c_{\alpha m}$ in Eq. (3.1.9) do not depend on time.

3.2 Phonons. Phonon-phonon interaction and relaxation

One of the important applications of the Born-Oppenheimer approximation is connected with the description of the crystal lattice. According to the adiabatic approximation we can describe the motion of atoms in crystal separately from the electronic motion. Thus we can introduce the potential energy of the crystal lattice depending only on nuclear coordinates (and electronic states). Having in mind that each atom in crystal has its equilibrium position, we can describe the potential energy of the crystal by the deviations from their equilibrium positions $\vec{R}^0$

$$\vec{u} = \vec{R} - \vec{R}^0 . \tag{3.2.1}$$

Then we can expand the potential of the lattice energy into series of the powers of $\vec{u}$ [50] and the Hamiltonian of the system of nuclei takes the form

$$H = \frac{1}{2}\sum_{lk\alpha} m_k \dot{u}_\alpha^2(lk) + \frac{1}{2}\sum_{lk\alpha l'k'\beta'} \Phi_{\alpha\beta}(l'k;lk')u_\alpha(lk)u_\beta(l'k')$$

$$\frac{1}{6}\sum_{lk\alpha}\sum_{l'k'\beta}\sum_{l''k''\gamma}\Phi_{\alpha\beta\gamma}(lk,l'k',l''k'')u_\alpha(lk)u_\beta(l'k')u_\gamma(l''k'')+\ldots \tag{3.2.2}$$

Quantities $u_\alpha(l,k)$ are Cartesian components of the displacement (3.2.1) of the k-th atom in the l-th unit cell, and M_k are masses of the k-th atom. Quantities Φ are the generalized force constants. Using the fact that the equilibrium configuration of atoms is the periodic structure, one can introduce normal (complex) coordinates

$$Q(\vec{k}, j)e^{i\vec{k}\vec{x}_l} , \tag{3.2.3}$$

diagonalizing harmonic (quadratic in displacements u) part of the Hamiltonian (3.2.2)

$$u_\alpha(lk) = \frac{1}{\sqrt{NM_k}}\sum_{\vec{k}j} e_\alpha(k \mid \vec{k}j)Q(\vec{k}j)\exp[i\vec{k}\vec{x}_l] . \tag{3.2.4}$$

Here $Q(\vec{k}j) = Q^*(-\vec{k}j), e_\alpha(k \mid \vec{k}j)$ is a vector of the polarization of the running wave (3.2.3). N is a number of cells in the crystal, and j numerates various branches of normal vibrations (acoustic, optical). In these coordinates the Hamiltonian (3.2.3) takes the form

$$H = \frac{1}{2}\sum_{\vec{k}j}\left[P(\vec{k}j)P(-\vec{k}j) + \omega_{\vec{k}j}^2 Q(\vec{k}j)Q(-\vec{k}j)\right]$$

$$+\frac{1}{3}\sum_{jj'j''\vec{k}\vec{k}'\vec{k}''}\Phi(\vec{k}j;\vec{k}'j';\vec{k}''j'')Q(\vec{k}j)Q(\vec{k}'j')Q(\vec{k}''j'')\Delta(\vec{k}+\vec{k}'+\vec{k}'')+\ldots \tag{3.2.5}$$

where

$$P(\vec{k}j) = \dot{Q}(\vec{k}j) \ ; \tag{3.2.6}$$

eigenfrequencies of the lattice are

$$\omega_{\vec{k}j} = \omega_{-\vec{k}j} \ , \tag{3.2.7}$$

anharmonic constants are determined by the force constants $\Phi_{\alpha\beta\gamma}, \ldots$, and function $\Delta(\vec{k})$ is different from zero only if $\vec{k} = 0$ or it is one of the vectors of the inversed lattice. Operators $Q(\vec{k}, j)$ and $P(\vec{k}, j)$ can be expressed through the operators of creation and annihilation of phonons

$$Q(\vec{k}j) = \sqrt{\frac{\hbar}{2\omega_{\vec{k}j}}}\left[a^{+}(-\vec{k}j) + a(\vec{k}j)\right] \ , \tag{3.2.8}$$

$$P(\vec{k}j) = i\sqrt{\frac{\hbar\omega_{\vec{k}j}}{2}}\left[a^{+}(-\vec{k}j) - a(\vec{k}j)\right] , \tag{3.2.9}$$

$$\left[a(\vec{k}j), a^{+}(\vec{k}'j')\right] = \delta_{\vec{k}\vec{k}'}\delta_{jj'} \ . \tag{3.2.10}$$

Using the above expressions one gets the harmonic part of the Hamiltonian (3.2.2)

$$H_0 = \frac{1}{2}\sum_{\vec{k}j}\hbar\omega_{\vec{k}j}\left[a^{+}(\vec{k}j)a(\vec{k}j) + a(\vec{k}j)a^{+}(\vec{k}j)\right] , \tag{3.2.11}$$

where

$$\varepsilon_{\vec{k}j} = \hbar\omega_{\vec{k}j} \ , \tag{3.2.12}$$

is the energy of the phonon, and

$$n(\vec{k}j) = a^{+}(\vec{k}j)a(\vec{k}j) \ , \tag{3.2.13}$$

is the operator of the phonon numbers. The anharmonic part of the Hamiltonian (3.2.2) contains products of the operators a and a^{+}, having more than two factors $aa^{+}a$, $a^{+}a^{+}a, \ldots, a^{+}aa^{+}a, \ldots$.

In the above consideration the displacement $u_a(lk)$, (3.2.4), is presented as a sum of traveling waves (3.2.3). Another possibility is to use the expansion over the standing waves. In this case, diagonalizing the harmonic part of the Hamiltonian (3.2.2) one gets

$$H = \frac{1}{2}\sum_{k}\left(p_k^2 + \omega_k^2 q_k^2\right) + \sum_{k_1,k_2,k_3} V_{k_1k_2k_3} q_{k_1} q_{k_2} q_{k_3} + \ldots \tag{3.2.14}$$

where k is the set of indices characterizing the vibration of the displacement.

Hamiltonian of the nuclei presented in the form (3.2.14) may describe not only phonons in the crystallic lattice, but the anharmonic vibrations in various electron-nuclear systems as well. It may, for example, describe the vibrational modes of huge molecules. In terms of the creation and annihilation operators a^+ and a, we get

$$q_k = \sqrt{\frac{\hbar}{2\omega_k}}(a_k^+ + a_k); \qquad p_k = i\sqrt{\frac{\hbar\omega_k}{2}}(a_k^+ - a_k) , \tag{3.2.15}$$

and the harmonic part of (3.2.14) has the form

$$H_0 = \sum_k \left(n_k + \frac{1}{2}\right)\hbar\omega_k; \qquad n_k = a_k^+ a_k . \tag{3.2.16}$$

The frequencies ω_k of the nuclear vibrations usually have a continuous spectrum and therefore the Hamiltonian (3.2.14) can be used to describe the irreversible relaxation process towards the equilibrium values of the phonon numbers. The transition between various phonon states is caused by the anharmonic terms in the Hamiltonian (3.2.14)

$$V_{k_1k_2k_3} q_{k_1} q_{k_2} q_{k_3} + \ldots \tag{3.2.17}$$

The master equation (2.2.28) describing the relaxation of phonons (or other kinds of harmonic vibrations) may be written in the form

$$\dot{p}(\{n\}) = -\sum_{\{n'\}} w_{\{n\}\{n'\}} [p(\{n\}) - p(\{n'\})] , \tag{3.2.18}$$

where $\{n\}$ is the manifold of various phonon numbers

$$\{n\} = (n_1, n_2, \ldots, n_k, \ldots) .$$

The example of the transition probability (per unit time) $w_{\{n\}\{n'\}}$ is

$$w_{n_{k_1} n_{k_2} n_{k_3} \to n_{k_1}+1, n_{k_2}-1, n_{k_3}-1}$$

$$= \frac{2\pi}{\hbar}\left|V_{k_1k_2k_3}\right|^2 \left[(q_{k_1})_{n_{k_1};n_{k_1}+1}(q_{k_2})_{n_{k_2};n_{k_2}-1}(q_{k_3})_{n_{k_3};n_{k_3}-1}\right]^2 \delta(\hbar\omega_{k_1} - \hbar\omega_{k_2} - \hbar\omega_{k_3})$$

$$= \frac{2\pi}{\hbar}\frac{\left|V_{k_1k_2k_3}\right|^2 \hbar^3}{8\omega_{k_1}\omega_{k_2}\omega_{k_3}}\left(n_{k_1}+1\right) n_{k_2} n_{k_3} \delta\left(\hbar\omega_{k_1} - \hbar\omega_{k_2} - \hbar\omega_{k_3}\right) . \tag{3.2.19}$$

The summation (3.2.18) over $\{n'\}$ is transformed into the integration over continuous frequencies ω_k.

3.3 Two-state electronic systems

In a number of applications, which we will consider below, it is a good approximation to consider only two states of the electronic system. In this case, which may be called the two-state approximation, we do not need the Born-Oppenheimer approximation; the Hamiltonian (3.1.1) can be treated exactly.

Let us consider an arbitrary two-state system. We introduce the effective spin operator $\vec{r}$ [51,21] defined by its three components, r_1, r_2, r_3. The operator $\vec{r}$ represents a vector not an ordinary geometrical space, but in some abstract space. The components of the vector $\vec{r}$ are determined by the commutation relations

$$\vec{r} \times \vec{r} = i\vec{r}; \quad [r_1,r_2] = ir_3; \quad [r_2,r_3] = ir_1; \quad [r_3,r_1] = ir_2 , \tag{3.3.1}$$

and the equations

$$r_1^2 = r_2^2 = r_3^2 = \frac{1}{4} . \tag{3.3.2}$$

Then in the representation in which r_3 is diagonal, the components r_i take the form

$$r_1 = \frac{1}{2}\begin{pmatrix} 0 & 1 \\ 1 & 0 \end{pmatrix}; \qquad r_2 = \frac{1}{2}\begin{pmatrix} 0 & -i \\ i & 0 \end{pmatrix}; \qquad r_3 = \frac{1}{2}\begin{pmatrix} 1 & 0 \\ 0 & -1 \end{pmatrix} . \tag{3.3.3}$$

It is easy to verify that these operators (3.3.3) satisfy the relations (3.3.1) and (3.3.2). Conventional 1/2 spin $\vec{s}$ is a particular case of the effective spin, when $(r_1,r_2,r_3) = (s_x,s_y,s_z)$ or $\vec{r} = \vec{s}$.

We will show that effective spin operators r_1, r_2 and r_3 can describe an arbitrary two-state system. For this purpose it is enough to see that the group of the linear Hermitian operators r_1, r_2, r_3 and the unit operator

$$I = \begin{pmatrix} 1 & 0 \\ 0 & 1 \end{pmatrix} , \tag{3.3.4}$$

has the property that every linear Hermitian operator A relating to a two-state system may be expressed in terms of these operators r_1, r_2, r_3, I

$$A = aI + br_1 + cr_2 + dr_3 = \begin{pmatrix} a + \frac{1}{2}d & \frac{1}{2}(b - ic) \\ \frac{1}{2}(b + ic) & a - \frac{1}{2}d \end{pmatrix} , \tag{3.3.5}$$

where a, b, c, and d are arbitrary real numbers. The right hand side of (3.3.5) is just the representation of an arbitrary linear Hermitian operator of a quantity characterized by two states. Thus the Hamiltonian of the system of nuclei and electrons (3.1.1), can be presented in the two-state approximation as follows

$$H=\sum_{k=1}^{N}\frac{p_k^2}{2M_k}+n_1U_1(q)+n_2U_2(q)+r_+V_{12}(q)+r_-V_{21}(q) \quad . \tag{3.3.6}$$

Here 1 and 2 designate two states of the electronic system, q is a set of nuclear variables, $q_1, q_2,\ldots,q_N$, and

$$n_1=\frac{1}{2}-r_3; \qquad n_2=\frac{1}{2}+r_3; \qquad r_\pm=r_1\pm ir_2 \quad . \tag{3.3.7}$$

When $r_3=-\frac{1}{2}$, $n_1=1$ and $n_2=0$, while at $r_3=\frac{1}{2}$, $n_2=1$ and $n_1=0$.

It is worthy to mention again that we do not use the adiabatic or Born-Oppenheimer approximation in deriving the Hamiltonian (3.3.6). Instead the two- state approximation is utilized. Only two states of the electronic system are taken into account and it is assumed that the influence of other states may be neglected.

Thus the Hamiltonian (3.3.6) describes the system with two electronic energy hypersurfaces

$$E_1=U_1(q_1,q_2,\ldots,q_N) \quad \text{and} \quad E_2=U_2(q_1,q_2,\ldots,q_N) \quad , \tag{3.3.8}$$

while $V(q_1,q_2,\ldots,q_N)$ is the interaction energy causing transition between these states. In order to carry out the analytical description it is widely accepted to use the model in which U_1 and U_2 are sets of potential energies of harmonic oscillators with identical frequencies and different equilibrium positions. We extend this model by taking into account anharmonic terms in Eq. (3.2.14) (with identical anharmonic constants $V_{kk_1k_2}$ in both electronic states)

$$U_1=J_1+\frac{1}{2}\sum_k\omega_k^2\left(q_k-q_{k1}^0\right)^2+\sum_{kk'k''}V_{kk'k''}\left(q_k-q_{k1}^0\right)\left(q_{k'}-q_{k'1}^0\right)\left(q_{k''}-q_{k''1}^0\right), \tag{3.3.9}$$

$$U_2=J_2+\frac{1}{2}\sum_k\omega_k^2\left(q_k-q_{k2}^0\right)^2+\sum_{kk'k''}V_{kk'k''}\left(q_k-q_{k2}^0\right)\left(q_{k'}-q_{k'2}^0\right)\left(q_{k''}-q_{k''2}^0\right). \tag{3.3.10}$$

These two potential energies may be presented as

$$U_1=J_1+U(\{q_k-q_{k1}^0\}); \quad U_2=J_2+U(\{q_k-q_{k2}^0\}) \quad , \tag{3.3.11}$$

where $U(\{q\})$ is the potential energy of the set of interacting anharmonic oscillators [Hamiltonian (3.2.5) or (3.2.14)].

The Hamiltonian (3.3.6) obtains the form (using the units $M_k = 1$)

$$H = \sum_k \frac{p_k^2}{2} + n_1 J_1 + n_2 J_2 + n_1 U\left(\{q_k - q_{k1}^0\}\right) + n_2 U\left(\{q_k - q_{k2}^0\}\right)$$
$$+ r_+ V_{12}(q) + r_- V_{21}(q) \quad . \tag{3.3.12}$$

For further application we will transform Hamiltonian (3.3.12) using the well known [52] unitary operator

$$S = \prod_k \exp\left[-\frac{i}{\hbar}(q_{k1}^0 n_1 + q_{k2}^0 n_2) p_k \right] , \tag{3.3.13}$$

where $\prod_k$ is a symbol of the products of k-factors. The transformed Hamiltonian, for which we shall preserve its former designation, takes the form

$$S^{-1} H S \rightarrow H = n_1 J_1 + n_2 J_2 + \frac{1}{2} \sum_k \left(p_k^2 + \omega_k^2 q_k^2\right)$$
$$+ \sum_{kk'k''} V_{kk'k''} q_k q_{k'} q_{k''} + \ldots + r_+ \Pi_1 V_{12} \Pi_2^+ + r_- \Pi_2 V_{21} \Pi_1^+ \quad , \tag{3.3.14}$$

where

$$\Pi_j = \prod_k \exp\left(\frac{i}{\hbar} q_{kj}^0 p_k \right) = \prod_k \exp[(-\eta_{jk}(a_k^+ - a_k)] \quad , \tag{3.3.15}$$

$$\eta_{ik} = \left(\frac{\omega_k}{2\hbar} \right)^{1/2} q_{kj}^0 \quad , \tag{3.3.16}$$

and a_k^+, a_k are creation and annihilation operators (3.2.15).

Further we will consider a simple case when

$$V_{12} = V_{21} = V \quad , \tag{3.3.17}$$

and does not depend on q (the Condon approximation). In this case the interaction energy takes the form

$$\hat{V} = r_+ V \Pi^+ + r_- V \Pi \quad , \tag{3.3.18}$$

where

$$\Pi = \prod_k \exp\left[-\eta_k(a_k^+ - a_k)\right], \quad \eta_k = \left(\frac{\omega_k}{2\hbar}\right)^{1/2}\left(q_{k1}^0 - q_{k2}^0\right) . \qquad (3.3.19)$$

For further considerations we will employ Hamiltonian (3.3.14) with the interaction energy (3.3.18)

$$H = n_1J_1 + n_2J_2 + \frac{1}{2}\sum_k\left(p_k^2 + \omega_k^2 q_k^2\right) + \sum_{kk'k''} V_{kk'k''}q_kq_{k'}q_{k''} + \ldots$$
$$+ r_+V\Pi^+ + r_-V\Pi . \qquad (3.3.20)$$

This Hamiltonian has the form of the Hamiltonian (2.1.2) where the Hamiltonian of the dynamic system E is that of a two level (two-state) system

$$E = n_1J_1 + n_2J_2 . \qquad (3.3.21)$$

The diagonalized Hamiltonian F of the dissipative system has the form

$$F = \frac{1}{2}\sum_k\left(p_k^2 + \omega_k^2 q_k^2\right) = \sum_k\left(n_k + \frac{1}{2}\right)\hbar\omega_k . \qquad (3.3.22)$$

The interaction Hamiltonian in the dissipative system is

$$G = \sum_{k,k',k''} V_{kk'k''}q_kq_{k'}q_{k''} + \ldots . \qquad (3.3.23)$$

This interaction is the cause in the relaxation of the dissipative system; this relaxation is described by the master equation (3.2.18). And finally, the interaction energy $\hat{V}$, which is the cause of the transitions between the levels of the dynamic system, has the form

$$\hat{V} = r_+V\prod_k \exp\left[\eta_k(a_k^+ - a_k)\right] + r_-V\prod_k \exp\left[-\eta_k(a_k^+ - a_k)\right] . \qquad (3.3.24)$$

In the general case (let us say, there is a different set of eigenfrequencies in two potential energies U_1 and U_2) the Hamiltonian (3.3.6) cannot be reduced to the form (3.3.20). Each electronic state has its own potential energy U_1 and U_2, (3.3.6). The Hamiltonian (3.3.6) may be taken in the form

$$H = r_3(J_2 - J_1) + \sum_k \frac{p_k^2}{2M_k} + \frac{1}{2}(\tilde{U}_1 + \tilde{U}_2) + r_3(\tilde{U}_1 - \tilde{U}_2)$$

$$+ r_+V_{12}(q) + r_-V_{21}(q) , \qquad (3.3.25)$$

where J_1 and J_2 are minima of the potential energies, a nonessential constant term $\frac{1}{2}(J_1 + J_2)$ is omitted, and $\tilde{U}_{1,2}$ are potential energies normalized to zero minima: $U_i = J_i + \tilde{U}_i$. Here the first term (in the right hand side) is the Hamiltonian E of the dynamic system, the second and third terms together present the Hamiltonian of the dissipative system

$$F + G = \sum_k \frac{p_k^2}{2M_k} + \frac{1}{2}(\tilde{U}_1 + \tilde{U}_2) \ , \qquad (3.3.26)$$

where F is the diagonalized part of this Hamiltonian and G the perturbation energy giving rise to the relaxation between eigenstates of F. The remaining terms on the right hand side of (3.3.25) represent the energy of interaction between the dynamic and dissipative system.

It should be mentioned that the total potential energy

$$U = \frac{J_1 + J_2}{2} + \frac{1}{2}(\tilde{U}_1 + \tilde{U}_2) \ , \qquad (3.3.27)$$

is infinitely large in comparison with the contribution of the two-level system $(\tilde{U}_1 - \tilde{U}_2)$, which is finite because it is connected with only one degree of freedom of the two level system, while the dissipative system has an infinite (in limit) number of degrees of freedom. As an example, let us consider the harmonic part of Eqs. (3.3.9) and (3.3.10), taking coordinates q such that

$$q_{k1}^0 = -q_{k2}^0 \ . \qquad (3.2.28)$$

Then

$$U = \frac{J_1 + J_2}{2} + \frac{1}{2}\sum \omega_k^2 q_k^2 + \frac{1}{2}\sum_k \omega_k^2 [(q_{k1}^0)^2 + (q_{k2}^0)^2] \ . \qquad (3.3.29)$$

The third term on the right hand side of Eq. (3.3.29) is finite, while the potential energy (a second term in Eq. (3.3.29)) tends to infinity when the number N of oscillators ω_k tends to infinity. If there is no singled out oscillator (like a local vibration), the third term in Eq. (3.3.29) can be finite, provided

$$q_{kj}^0 \propto \frac{1}{\sqrt{N}} \ . \qquad (3.3.30)$$

3.4 Radiationless transitions. Basic models of electron and energy transfer

In the huge and medium-size molecules with a large number of molecular vibrations (vibrations of the molecule's atoms) radiationless transitions can take place (without the electromagnetic radiation) between electronic states of the molecule. These transitions can be described by the Hamiltonian (3.3.20). Such a description (without anharmonic terms) was presented in the pioneering work of Lin [53]. It had a remarkable development in the works of Jortner, Bixon, Englman and others [54-57]. An extensive updated bibliography is given in the review of Bixon and Jortner [58], devoted to electron transfer. As a matter of fact electron and energy transfer processes may be considered as a specific case of radiationless processes, and they are described by the same formalism.

Electron transfer is one of the important processes going on in the condensed media, in general, and in huge molecules and biomolecular systems. Pioneering works describing the electron transfer are those of Marcus [59] and Levich [60]. An important development of the theory of electron transfer and application is presented in the review by Bixon and Jortner [58].

We present here basic models of the electron and energy transfer.

Electron transfer. The quantum-theoretical description of the electron transfer was given by Levich [60]. Let ε_1 and ε_2 be the electronic terms, each of which correspond to different localizations ξ_1 and ξ_2 of the electron (Fig. 2). Then the transition from state ε_1 and state ε_2 will be accompanied by the electron transfer from the point ξ_1 (or to be exact, from the vicinity of this point) to point ξ_2.

In the adiabatic approximation the electronic energies essentially depend on the nuclear configuration and electran transfer is accompanied by the change of the configuration. In the two-state model of section 3.3 two electronic states correspond to two localizations described by the potential energies $U_1(q)$ and $U_2(q)$ in Eq. (3.3.6) while $r_3 = \frac{1}{2}$ and $-\frac{1}{2}$ describe two positions of nuclei (in the vicinity of minima of U_1 and U_2). Why is it necessary that the change of the nuclear configuration should occur? Let us look at Fig. 2a, which corresponds to the configuration of nuclei $q = q_1$ (which may be a set of positions of nuclei). In this configuration there is a large gap between energies ε_1 and ε_2 (in the two-state model $U_1(q_1)$ is essentially different from $U_2(q_1)$). At a large gap

$$|\varepsilon_1 - \varepsilon_2| >> |V| \quad , \tag{3.4.1}$$

the tunneling between these states has a very low probability. But when the configuration of nuclei changes from q_1 to q_0, Fig. 2c, then the energy gap is zero and the tunneling will have its maximum value. Thus nuclear motion stimulates the electron transfer. In the condensed medium (or the biomolecule) the initial and final states (the electron before and after the transfer) are characterized by the electron energy hypersurfaces

$$U_1 = \varepsilon_1(q_1, q_2, \ldots, q_N); \quad U_2 = \varepsilon_2(q_1, q_2, \ldots, q_N) \tag{3.4.2}$$

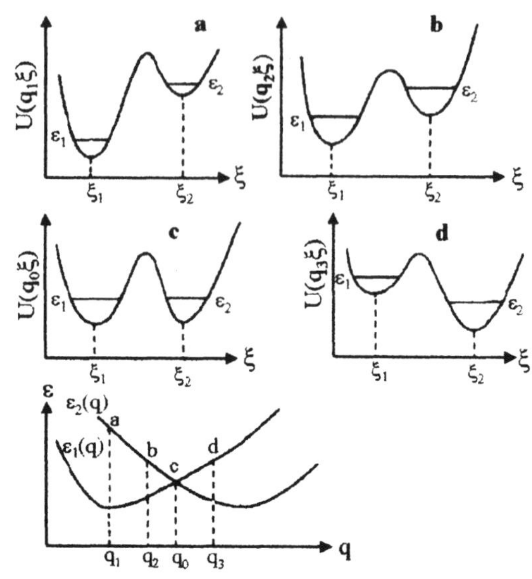

Fig. 2. Potential energies of electron $U(q,\xi)$ and energy levels ε_1 and ε_2 as a function of nuclear coordinates. Curves a, b, c, and d correspond to various values of the nuclear coordinate q. The lower curve describes dependence of the electronic energies ε_1 and ε_2 on nuclear coordinates q.

Transitions between these hypersurfaces imply the electron transfer. Using the potential energies (3.3.9) and (3.3.10) we come to the Hamiltonian (3.3.20) for a description of the electron transfer

$$H = n_1 J + n_2 J + \frac{1}{2}\sum_k \left(p_k^2 + \omega_k^2 q_k^2\right) + \sum_{kk'k''} V_{kk'k''} q_k q_{k'} q_{k''} + r_+ V\Pi^+ + r_- V\Pi^- \quad , \quad (3.4.3)$$

where Π is defined by Eq. (3.3.19).

Energy transfer. The energy transfer between atoms (molecules) embedded in condensed media or energy transfer between different subunits of a macro-molecular framework is of vital importance to diverse fields of research such as sensitized luminescence and photosynthesis. Starting from pioneering works of Förster [61] and Dexter [62] a lot of theoretical and experimental works devoted to this problem have been carried out. However, the aim of this section is to provide basic physical models of corresponding processes.

We want to present the theoretical framework for the description of energy transfer in condensed media. The Hamiltonian describing energy transfer between two atoms (molecules) embedded in a condensed medium was discussed by Soules and Duke [63], Rackovsky and Silbey [64], and Abram and Silbey [65]. We will consider the simplified model, taking into account only two levels of each molecule. We neglect transitions between energy levels of each molecule and take into account only the energy transfer process between these molecules caused by the intermolecular interaction V (see Fig. 3). In this case the two-molecule system may be in two possible states. In state 1, the first molecule is in the excited state, while the second molecule is in the ground state. In state 2, the second molecule is in the excited and the first molecule in the ground state. Respectively, the eigenenergies of the two-molecule system are E_1 and E_2.

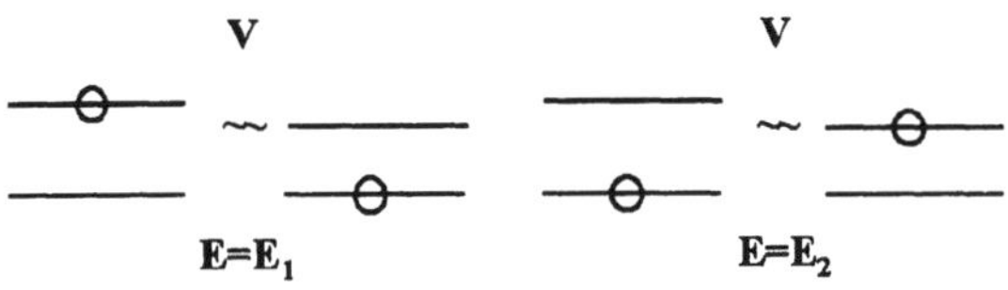

Fig. 3. Energy levels of two molecules coupled by the interaction energy.

Using the formalism of Section 3, with travelling wave expansion (3.2.3) of the displacement in the system of two molecules we obtain the Hamiltonian

$$H = n_1 J + n_2 J + \sum_{\vec q} \hbar\omega_{\vec q}\left(b_{\vec q}^+ b_{\vec q} + \frac{1}{2}\right) + G + r_+ V\Pi^+ + r_- V\Pi \quad . \tag{3.4.4}$$

Here G is the anharmonic part of the potential energy, containing the sum of the terms of the $b_q b_{q'}^+ b_{q''}$ type,

$$\Pi = \prod_{\vec q} \exp\left[-\frac{1}{\hbar}\eta(\vec q)(b_{\vec q} - b_{-\vec q}^+)\right] \quad , \tag{3.4.5}$$

$$\eta(\vec q) = \eta_1(\vec q)e^{i\vec q\vec R_1} - \eta_2(\vec q)e^{i\vec q\vec R_2} \quad , \tag{3.4.6}$$

where $\vec R_1$ and $\vec R_2$ are coordinates of two molecules in the dimer (two molecules)

$$\eta_1(\vec q) = G_1^e(\vec q) - G_1^g(\vec q); \qquad \eta_2(\vec q) = G_2^e(\vec q) - G_2^g(\vec q) \quad . \tag{3.4.7}$$

Quantities G_i^e and G_i^g are the constants of the electron-phonon coupling in the excited and ground state of the i-th molecule respectively. These constants are connected with the equilibrium oscillator coordinates q_{k1}^0 and q_{k2}^0 appearing in Eqs. (3.3.9 – 3.3.12)

$$Q_1^0(\vec q) = \sqrt{\frac{\hbar}{2\omega_q}}\left[G_1^e e^{i\vec q\vec R_1} + G_2^g e^{i\vec q\vec R_2}\right] \quad , \tag{3.4.8}$$

$$Q_2^0(\vec{q}) = \sqrt{\frac{\hbar}{2\omega_q}}\left[G_1^g e^{i\vec{q}\vec{R}_1} + G_2^e e^{i\vec{q}\vec{R}_2}\right] . \tag{3.4.9}$$

Thus we see that Hamiltonian (3.4.4) is isomorphous to that of (3.4.3) describing the electron transfer. Of course the Hamiltonian (3.4.4), as well as (3.4.3), are isomorphous to the general form of the Hamiltonian (3.3.20), describing radiationless transitions in general.

Temperature dependence of the energy transfer rates in various temperature regions, starting from $T = 0$ till very large temperature $T \to \infty$, is given in paper [66].

3.5 Tunneling in the condensed media

The Born-Oppenheimer approximation (see section 3.1) gives the possibility to understand the essence of chemical transformations and other processes, including those occurring in condensed media. According to this approximation, the motion of nuclei, atoms and molecules may be described by the effective potential energy U_α averaged over electronic eigenstate α (see (3.1.8)).

We consider here a quite general model [67] which may be appropriate for the description of proton transfer, nuclear group transfer, electron transfer accompanied by the transfer of nuclear groups, and other rate processes in condensed media. These transitions may occur between two intersecting potential energies belonging to different electronic states (Fig. 4). These processes are called non-adiabatic transitions.

On the other hand, all above processes may have taken place on the same electronic state: transitions between two minima of the same potential energy (belonging to the same electronic state (Fig. 5).

These processes are called adiabatic processes. In the Born-Oppenheimer the Hamiltonian describing adiabatic transition processes may be presented in the form

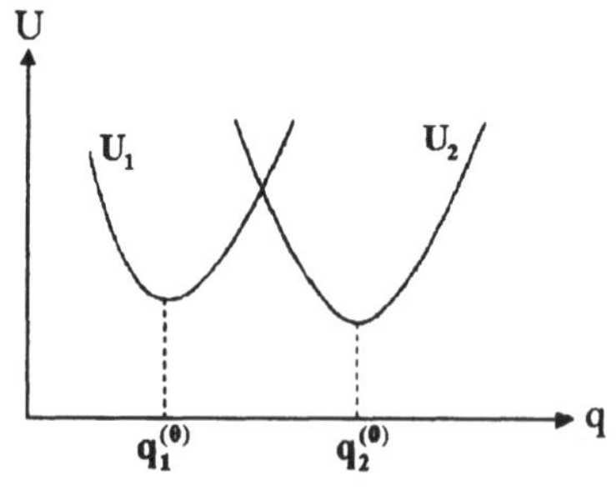

Fig. 4. Two intersecting potential energy curves (hypersurfaces). Non-adiabatic case.

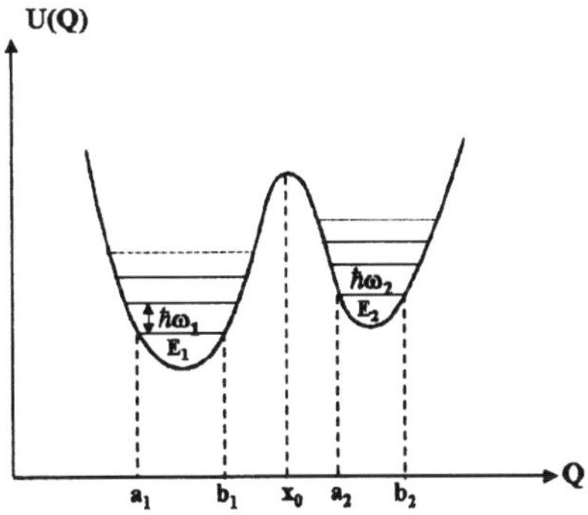

Fig. 5. The adiabatic energy curve (hypersurface) with two potential well. E_1 and E_2 ground states in these wells. $\hbar\omega_1$ and $\hbar\omega_2$ are energy level differences in the first and second well respectively. a_1, b_1, a_2 and b_2 are the turning points.

$$H = T + U(Q) + \frac{1}{2}\sum_k \left(p_k^2 + \omega_k^2 q_k^2\right) + G(q) - \sum_k A_k(Q) q_k \quad . \tag{3.5.1}$$

Here Q is the set of coordinates of the singled out modes subsystem interacting with its surrounding - dissipative system, thermal bath. T and U are kinetic and potential energies of the nuclear subsystem; the third system in the Hamiltonian (3.5.1) describes the condensed medium in the harmonic approximation, while G(q) describes its anharmonic part. The last term is the interaction energy between the nuclear subsystem and the condensed medium, the only assumption about this interaction is that it is linear in the coordinates q_k, i.e., the excitations of the surrounding condensed medium are small enough.

The potential energy U is supposed to have two minima (Fig. 5) corresponding to two (quasi) stable configurations of the nuclear subsystem. The condition of the (quasi) stability of the configurations is that they are divided by a barrier which is sufficiently large (high and wide), such that the duration of transition between two minima is much smaller than the period of the oscillation in each of the wells. The transition from one configuration (Fig. 5) (potential well 1) to another (potential well 2) may occur in two ways: (1) tunneling through the potential barrier, and (2) overcoming the potential barrier by thermal fluctuations induced by the vibrations of the thermal bath. Here we will explore the first possibility, i.e. the tunneling.

We would assume that only two low-lying levels of the subsystem are essential, i.e. two levels E_1 and E_2 in each well (with a sufficiently large potential barrier) and it is possible to neglect excitations to other levels. It may mean that the temperature of the thermal bath is small enough

$$k_B T \ll \hbar\omega_0 \quad , \tag{3.5.2}$$

where $\hbar\omega_0$ is the characteristic energy difference between the next higher level of the first (second) well and E_1 (E_2). In this case we are dealing with a two-state system (see section (3.3)). (The finite temperature case has been considered by Dekker [68,69].)

The Hamiltonian of the two-state nuclear system without thermal bath, may be presented in the form (3.3.5)

$$E = n_1 E_1 + n_2 E_2 + r_+ V_{12} + r_- V_{21} \ , \tag{3.5.3}$$

where operators n_1, n_2, $r_\pm$ are defined by Eqs. (3.3.7), (3.3.1) and (3.3.2), and V_{12}, V_{21} are the matrix elements of the effective perturbation energy, causing the tunneling between states 1 and 2. It has been shown [67,18] that the effective perturbation energy matrix elements for the one-dimensional case in the mixed semiclassical, ground state approximation has the form (see Fig. 3.4)

$$V_{12} = V_{21} = \frac{1}{2}\hbar(\omega_1\omega_2/\pi e)^{1/2}\exp[-\hbar^{-1}\int_{b_1}^{a_2}|p|\,dQ] \ , \tag{3.5.4}$$

where ω_1 and ω_2 are frequencies characterizing wells 1 and 2. For ground states E_1 and E_2 the parabolic approximation is a good one. Therefore the potential energies may be described by frequencies ω_1 and ω_2. In Eq. (3.5.4)

$$p = \sqrt{2M(E-U)} \ , \tag{3.5.5}$$

where $E \approx E_1 \approx E_2$. In the case when E_1 substantially differs from E_2, the tunneling is practically unachievable.

In the general case it is reasonable to assume that the matrix elements V have the form

$$V = \hbar\omega_0 e^{-\sigma} \ , \tag{3.5.6}$$

where the effective parameter ω_0 has the order of magnitude of that appearing in Eq. (3.5.4), $(\omega_1 \approx \omega_2 \approx \omega_0)$, and the parameter σ is assumed to be much larger than unity

$$\sigma >> 1 \ . \tag{3.5.7}$$

Now let us consider the time-dependence of the density matrix of the system described by the two-state Hamiltonian (3.5.3). In the site representation (in which the system is either in the left or right potential well) the time dependence of the density matrix has the form

$$\dot{\sigma}_{11} = -\dot{\sigma}_{22} = -i\hbar^{-1}V(\sigma_{21} - \sigma_{12}) \ , \tag{3.5.8}$$

$$\dot{\sigma}_{11} = \dot{\sigma}_{21} = -i\omega_{12}\sigma_{12} - i\hbar^{-1}V(\sigma_{22} - \sigma_{11}) \ , \tag{3.5.9}$$

where ω_{12} = $(E_1\text{-}E_2)/\hbar$. Performing simple manipulations we get for the population difference

$$n = \sigma_{11} - \sigma_{22} \ ; \quad \sigma_{11} + \sigma_{22} = 1 \ , \tag{3.5.10}$$

the equation

$$\ddot{n} + \Omega^2 \dot{n} = 0 \ , \tag{3.5.11}$$

where

$$\Omega = \hbar^{-1}(4V^2 + \hbar^2\omega_{12}^2)^{1/2} \ . \tag{3.5.12}$$

Equation (3.5.11) has a general solution

$$n = A \cos \Omega t + B \sin \Omega t + C \ . \tag{3.5.13}$$

If at the initial time $t = t_0$ the elements of the density matrix are

$$\sigma_{12}(t_0) = \sigma^*_{21}(t_0) = 0 \ ; \quad \sigma_{11}(t_0) = 1; \ \ \sigma_{22}(t_0) = 0 \ , \tag{3.5.14}$$

then

$$n = \left[4V^2\hbar^{-2}\cos(\Omega(t - t_0)) + \omega_{12}^2\right]/\Omega^{-2} \ , \tag{3.5.15}$$

while

$$\sigma_{11}(t - t_0) = \frac{1}{2} + \frac{1}{2}\frac{4V^2/\hbar^2 \cos\Omega(t - t_0) + \omega_{12}^2}{4V^2/\hbar^2 + \omega_{12}^2} \ . \tag{3.5.16}$$

In the case of the symmetrical two-well potential energy $\omega_{12} = 0$

$$\sigma_{11}(t - t_0) = \frac{1}{2}(1 + \cos\Omega_0(t - t_0)); \qquad \Omega_0 = \frac{2V}{\hbar} \ . \tag{3.5.17}$$

Eqs. (3.5.15 and 3.5.16) describe so-called quantum beats between two potential wells.

Now, taking into account the interaction with the dissipative system (the phonon bath) and taking into account only two levels E_1 and E_2 of the nuclear subsystem, the Hamiltonian (3.5.1) of the whole system may be presented in the form

$$H = n_1 H_{11} + n_2 H_{22} + r_+ H_{12} + r_- H_{21} \ , \tag{3.5.18}$$

where the matrix elements are taken with the aid of the eigenfunctions $\psi_1(Q)$ and $\psi_2(Q)$, with eigenvalues E_1 and E_2. Thus, the effective Hamiltonian of the whole system (including the anharmonic phonon bath) takes the form

$$H = n_1 E_1 + n_2 E_2 + \frac{1}{2}\sum_k \left(p_k^2 + \omega_k^2 q_k^2\right) + G(q) - n_1 \sum_k \omega_k^2 q_{k1}^0 q_k$$

$$- n_2 \sum_k \omega_k q_{k2}^0 q_k + r_+\left[V_{12} + \sum_k V_{12k} q_q\right] + r_-\left[V_{21} + \sum_k V_{21k} q_k\right] , \quad (3.5.19)$$

where

$$\omega_k^2 q_{k1}^0 = \int \psi_1^* A_k(Q) \psi_1 dQ , \quad (3.5.20)$$

$$\omega_k^2 q_{k2}^0 = \int \psi_2^* A_k(Q) \psi_2 dQ , \quad (3.5.21)$$

$$V_{12k} = V_{21k}^* = -\int \psi_1^* A_k(Q) \psi_2 dQ , \quad (3.5.22)$$

and $Q = \{Q_1,\ldots,Q_n\}$.

Apart from the terms V_{12k}, V_{2k} (which are neglected in the Condon approximation) it is easy to see that the Hamiltonian (3.5.19) is identical with Hamiltonian (3.3.6) with U_1 and U_2 determined by Eqs. (3.3.9) and (3.3.10). It means, also, that Hamiltonian (3.5.19) is isomorphous with those describing radiationless transitions in general and the electron, (3.4.3), and energy, (3.4.4), transfer, in particular.

3.6 Equations of motion of the two-state electronic (nuclear) system interacting with vibrations of the medium

The result of the preceding sections shows that such versatile processes as radiationless transitions in huge molecules, the electron transfer in condensed media, the energy transfer and others can be presented by the Hamiltonian (3.3.20). This Hamiltonian describes the two-level system

$$E = n_1 J_1 + n_2 J_2 , \quad (3.6.1)$$

interacting with the system of anharmonic oscillators (3.3.22) and (3.3.23)

$$F + G = \frac{1}{2}\sum \left(p_k^2 + \omega_k^2 q_k^2\right) + \sum_{kk'k''} V_{kk'k''} q_k q_{k'} q_{k''} + \ldots , \quad (3.6.2)$$

with the specific kind of the interaction energy, (3.3.18),

$$\hat{V} = r_+ \prod_k \exp\left[\eta_k (a_k^+ - a_k)\right] + r_- V \prod_k \exp\left[-\eta_k (a_k^+ - a_k)\right] , \qquad (3.6.3)$$

where

$$\eta_k = \left(\frac{\omega_k}{2\hbar}\right)^{1/2} \left(q_{k1}^0 - q_{k2}^0\right) , \qquad (3.6.4)$$

and q_{k1}^0, q_{k2}^0 are the equilibrium positions of two electronic hypersurfaces (3.3.9 – 3.3.11).

The time development of the above mentioned processes, such as electron transfer, etc., is described by the equations (2.7.31, 2) of the two state system

$$\dot{\sigma}_{11} = -\dot{\sigma}_{22} = -i\hbar^{-1}\left(\overline{V}_{12}^{ef}\sigma_{21} - \sigma_{12}\overline{V}_{21}^{ef}\right) - \left(w_{12}\sigma_{11} - w_{21}\sigma_{22}\right) , \qquad (3.6.5)$$

$$\dot{\sigma}_{21} = -\dot{\sigma}_{21}^* = -i\omega_{12}\sigma_{12} - i\hbar^{-1}\overline{V}_{12}^{ef}\left(\sigma_{22} - \sigma_{11}\right) - R\sigma_{12} + R^*\sigma_{21} . \qquad (3.6.6)$$

Here, (2.4.13)

$$\overline{V}^{ef} = \overline{V} + i\hbar^{-1} \int_0^\infty \langle [V(\tau), G(0)] \rangle d\tau , \qquad (3.6.7)$$

where G is the anharmonic part of the Hamiltonian, (3.6.2), the averaging $\overline{V}$, <...> is performed over the states of the dissipative system and w_{12}, w_{21} are determined by Eq. (2.7.9), while the quantity R is determined by Eq. (2.7.33). In the specific case of the interaction (3.6.3) and using the definition of $\tilde{V}$, (2.4.4), we get expressions for the rate coefficients

$$w_{12} = V^2\hbar^{-2} \int_{-\infty}^{\infty} e^{-i\omega_{12}\tau} d\tau$$

$$\times \left\langle \Pi_k \left\{ \exp[\eta_k (a_k^+(\tau) - a_k(\tau))] \exp[-\eta_k (a_k^+ - a_k)] - \exp[-\eta_k^2 (a_k^+ - a_k)^2] \right\} \right\rangle , (3.6.8)$$

$$w_{21} = V^2\hbar^{-2} \int_{-\infty}^{\infty} e^{-i\omega_{21}\tau} d\tau$$

$$\left\langle \Pi_k \left\{ \exp[-\eta_k (a_k^+(\tau) - a_k(\tau))] \exp[\eta_k (a_k^+ - a_k)] - \exp[-\eta_k^2 (a_k^+ - a_k)^2] \right\} \right\rangle , \quad (3.6.9)$$

$$R = 2V^2\hbar^{-2}\int_0^\infty e^{-i\omega 21\tau}d\tau$$

$$\left\langle \operatorname{Re}\prod_k\left\{\exp[\eta_k(a_k^+(\tau)-a_k(\tau))]\exp[-\eta_k(a_k^+-a_k)]-\exp[-\eta_k^2(a_k^+-a_k)^2]\right\}\right\rangle \quad (3.6.10)$$

It has been mentioned at the end of Section 2, Eq. (3.3.30), that $q_{kj}^0 \propto N^{-1/2}$, and the same refers to η_k.

In this case (in the limit $N \to \infty$) the product in the integrand (3.6.8) may be transformed as follows (see Eq. (3.3.19)

$$<\Pi^+(\tau)\Pi(0)>=$$

$$\prod_k\left\langle[(1+\eta_k(a_k^+(\tau)-a_k(\tau))+\frac{1}{2}\eta_k^2(a_k^+(\tau)-a_k(\tau))^2][1-\eta_k(a_k^+-a_k)+\frac{1}{2}\eta_k^2(a_k-a_k)^2]\right\rangle$$

$$=\exp\left\{-\sum_k\left[\eta_k^2(2\bar{n}_k+1)-\eta_k^2(\bar{n}_k e^{i\omega_k\tau}+(\bar{n}_k+1)e^{-i\omega_k\tau})\right]\right\} \quad . \quad (3.6.11)$$

Here we have assumed that in the equilibrium state of the dissipative system

$$\left\langle a_k^+a_k^+\right\rangle=\left\langle a_k a_k\right\rangle=0 \quad . \quad (3.6.12)$$

We have also used the explicit time dependence of operators a^+,a

$$a_k^+(t)=a_k^+e^{i\omega_k t}, \quad a_k(t)=a_k e^{-i\omega_k t} \quad . \quad (3.6.13)$$

Using Eq. (3.6.11) and Eqs. (3.6.8), (3.6.9) and (3.6.10), we get

$$w_{12}=\frac{V^2}{\hbar^2}e^{-F(0)}\int_{-\infty}^{\infty}\left(e^{F(\tau)}-1\right)e^{i\omega 12\tau}d\tau \quad , \quad (3.6.14)$$

$$w_{21}=\frac{V^2}{\hbar^2}e^{-F(0)}\int_{-\infty}^{\infty}\left(e^{F(\tau)}-1\right)e^{i\omega 21\tau}d\tau \quad , \quad (3.6.15)$$

$$R=\frac{2V^2}{\hbar^2}e^{-F(0)}\int_0^\infty e^{i\omega 21\tau}\operatorname{Re}\left(e^{F(\tau)}-1\right)d\tau \quad , \quad (3.6.16)$$

and

$$\overline{V}^{ef} = Ve^{-\frac{1}{2}F(0)} + (i\hbar)^{-1}\int_0^{\infty}\langle[\overline{V}(\tau), G(0)]\rangle d\tau \ . \tag{3.6.17}$$

Here

$$F(\tau) = \sum_k \eta_k^2\left[\overline{n}_k e^{-i\omega_k\tau} + (\overline{n}_k + 1)e^{i\omega_k\tau}\right], \tag{3.6.18}$$

$\overline{n}_k$ are equilibrium phonon occupation numbers and η_k is defined by Eq. (3.6.4). In the derivation of these equations the smallness of $\eta_k \propto N^{-\frac{1}{2}}$ has been employed, however, the same relations can be achieved when the averaging is performed over the thermodynamic equilibrium.

The transition probabilities w_{12}, w_{21} and R contain the compensating factor -1 in the integrands (3.6.14 – 3.6.16). This factor eliminates the singularities in the expressions (3.6.14 – 3.6.16) when $\omega_{12} = 0$.

It is worthwhile to mention that equations (3.6.5) and (3.6.6), apart from the term with R^*, coincide with the Bloch equations [26] for spin $\frac{1}{2}$ in the external field, $\overline{V}^{ef}$ representing the interaction energy with this field. The term with R^* is essential in the case of almost degenerate levels J_1 and J_2 when conditions of the (2.7.4) type are not fulfilled. The meaning of the terms with R^* in equation (3.6.6) is the transformation $\sigma_{12} \rightleftarrows \sigma_{21}$ while in the usual Bloch equations such transformation does not exist. In the case

$$|\omega_{12}| >> |V_{12}^{ef}|/\hbar, w_{12}, w_{21}, |R| \ , \tag{3.6.19}$$

which corresponds to conditions (2.7.4) and (2.7.15), the equation (3.6.5) obtains the form

$$\dot{\sigma}_{11} = -\dot{\sigma}_{22} = -(w_{12}\sigma_{11} - w_{21}\sigma_{22}) \ , \tag{3.6.20}$$

which coincides with the master equation for the two-level system. In this case, Eq. (3.6.6) can be approximated by equation (2.7.16) for the two level system. In the latter equation the transformation $\sigma_{12} \rightleftarrows \sigma_{21}$ is neglected.

3.7 Calculation of rate coefficients

The transition probabilities (3.6.14), (3.6.15) and parameter R, (3.6.16), are rate coefficients of the equations (3.6.5) and (3.6.6). Now we will try to perform a general analysis of these quantities in various regions of their parameters. As we know from the general theory (Chapter I, section 7) the transition probabilities per unit time can be determined only for the systems with continuous spectra. More than that, it has been shown (section 1.8) that the continuous spectrum of energies is a necessary condition of the irreversible motion in general. It means that the sum (3.6.18) can be transformed into the Fourier integral

$$F(\tau) = \sum_k \eta_k^2 \left[\bar{n}_k e^{-i\omega_k \tau} + (\bar{n}_k + 1) e^{i\omega_k \tau}\right] = \int_{-a}^{a} \rho(\omega) e^{-i\omega\tau} d\omega \ . \tag{3.7.1}$$

Here we have introduced the following designations

$$\rho(\omega) = f(\omega)\eta^2(\omega)\,|\,n(\omega)\,| \ , \tag{3.7.2}$$

where $f(\omega)$ is the frequency distribution function

$$\sum_k \rightarrow \int_0^a f(\omega) d\omega \ , \tag{3.7.3}$$

a is the maximum frequency of the vibrational spectrum,

$$\eta(\omega) = \eta_k(\omega) \ ,$$

and $n(\omega)$ is the average number of phonons with frequency ω; in the case of thermal equilibrium

$$n(\omega) = \left[e^{\hbar\omega/k_B T} - 1\right]^{-1} \ , \qquad (\omega > 0) \ . \tag{3.7.4}$$

For negative frequencies $-\omega$ we define $(\omega > 0)$

$$f(-\omega) = f(\omega) \ ; \qquad \eta^2(-\omega) = \eta^2(\omega) \ ; \qquad n(-\omega) = n(\omega) + 1 \ . \tag{3.7.5}$$

In the case of thermal equilibrium, (3.7.4), we have for negative frequencies

$$|\,n(\omega)\,| = |\,n(-\,|\,\omega\,|)\,| = \left|\frac{1}{e^{-\hbar|\omega|/kT} - 1}\right| = n(|\,\omega\,|) + 1 \ , \tag{3.7.6}$$

Thus in the general case, taking into account definition (3.7.5), or the equilibrium case (3.7.6), we can use the distribution (3.7.2).

The calculation of transition probabilities w_{12} and w_{21}, (3.6.14), (3.6.15), may be reduced to the calculation of the integral

$$I(\lambda) = \int_{-\infty}^{\infty} e^{\psi(\tau)} d\tau \ , \tag{3.7.7}$$

where

$$\psi(\tau) = F(\tau) + i\lambda\tau = \int_{-a}^{a} \rho(\omega) e^{-i\omega\tau} d\omega + i\lambda\tau \ . \tag{3.7.8}$$

Integral (3.7.7) has a typical form suitable for the calculation by the saddle point method (see, e.g. [71]). According to this method, the exponent in (3.7.7) should contain a large factor. Later on we will clarify what this means. If this factor tends to infinity, the integral (3.7.7) has the asymptotic representation

$$I(\lambda) = \sum_{z_o} e^{\psi(z_o)} \left[2\pi/(-\psi''(z_o))\right]^{1/2} \ , \tag{3.7.9}$$

where z_o are all saddle points along the integration contour, (3.7.7), shifted to the complex region. The saddle points satisfy the equations

$$\psi'(z_o) = 0 \ . \tag{3.7.10}$$

The saddle point method is based on an expansion of the exponent $\psi(\tau)$ in the vicinity of the saddle point

$$\psi(z) = \psi(z_o) + \frac{1}{2}\psi''(z_o)(z - z_o)^2 \ , \tag{3.7.11}$$

up to terms of second order in $(z\text{-}z_o)$. This means that in the region where the second term in (3.7.11) gives an essential contribution to the integral $I(\lambda)$

$$| z - z_o | \lesssim | \psi''(z_o) |^{-2} \ , \tag{3.7.12}$$

the next terms of the expansion of the function $\psi(z_o)$ should be small:

$$| \psi'''(z_o)(z - z_o)^3 | \sim | \psi'''(z_o) | / | \psi''(z_o) |^{3/2} \sim \frac{1}{\left| \int_{-a}^{a} \rho(\omega) d\omega \right|^{1/2}} << 1 \ , \tag{3.7.13}$$

$$|\psi^{IV}(z_o)(z-z_o)^4| \sim |\psi^{IV}(z_o)|/|\psi''(z_o)|^2 \sim \frac{1}{\left|\int_{-a}^{a}\rho(\omega)d\omega\right|} << 1 \ . \tag{3.7.14}$$

Just these conditions determine a large dimensionless factor which is necessary for the application of the saddle point method

$$\int_{-a}^{a}\rho(\omega)d\omega >> 1 \ . \tag{3.7.15}$$

Now we will analyze integral $I(\lambda)$ in various limiting case [72,18]. In the complex region

$$z_o = x_o + iy_o \ , \tag{3.7.16}$$

Eq. (3.7.10) takes the form

$$-\int_{-a}^{a}\omega\rho(\omega)e^{-i\omega x_o}e^{\omega y_o} + \lambda = 0 \ , \tag{3.7.17}$$

or

$$\int_{-a}^{a}\omega\rho(\omega)\sin\omega x_o e^{\omega y_o}d\omega = 0 \ , \tag{3.7.18}$$

$$\int_{-a}^{a}\omega\rho(\omega)\cos\omega x_o e^{\omega y_o}d\omega = \lambda \ . \tag{3.7.19}$$

We will consider the case corresponding to small $|y_o|$

$$a|y_o| << 1 \ . \tag{3.7.20}$$

In this approximation we find from (3.7.18) and (3.7.19) the equations determining x_o and y_o

$$\int_{-a}^{a}\omega\rho(\omega)\sin\omega x_o d\omega = 0 \ , \tag{3.7.21}$$

$$y_o = \frac{\lambda - \int_{-a}^{a} \rho(\omega)\omega \cos \omega x_o d\omega}{\int_{-a}^{a} \rho(\omega)\omega^2 \cos \omega x_o d\omega} . \tag{3.7.22}$$

Formulae (3.7.8), (3.7.9), (3.7.21) and (3.7.22) determine the integral (3.7.7) in the saddle point approximation and the approximation (3.7.20). Generally speaking there are an infinite number of saddle points x_o, y_o that give contributions to the integral (3.7.9). Apart from exceptional cases, the summation over an infinite number of saddle points z_o cannot be performed in an analytical way. This is why it is important to find a condition under which the main contribution to the integral $I(\lambda)$, (3.7.9), comes just from one saddle point (particularly $x_o = 0$). It is easy to show that the condition of one saddle point $x_o = 0$ takes the form

$$\exp[\psi(x_o \neq 0) - \psi(x_o = 0)] \propto \exp\left[\int_{-a}^{a} \rho(\omega)(1 - \cos \omega x_o) d\omega\right] >> 1 . \tag{3.7.23}$$

This condition can be written in the form

$$\int_{-a}^{a} \rho(\omega)(1 - \cos \omega x_o) d\omega >> 1 , \tag{3.7.24}$$

where x_o is a non-zero root of Eq. (3.7.21). This condition is consistent with the inequality (3.7.15) determining the applicability of the saddle point method. The condition of our approximation (3.7.20) can be deduced from Eq. (3.7.22)

$$\left|\lambda - \int_{-a}^{a} \rho(\omega)\omega d\omega\right| << \frac{1}{a} \int_{-a}^{a} \rho(\omega)\omega^2 d\omega . \tag{3.7.25}$$

Provided all these conditions are fulfilled, the expression for the transition probability obtains the form [see (3.6.14), (3.7.7), (3.7.8), (3.7.9) and (3.7.22)]

$$w_{12} = \frac{V^2}{\hbar^2} [2\pi / \int_{-a}^{a} \rho(\omega)\omega^2 d\omega]^{\frac{1}{2}} \exp[- \frac{(\omega_{12} - \int_{-a}^{a} \rho(\omega)\omega d\omega)^2}{2\int_{-a}^{a} \rho(\omega)\omega^2 d\omega}] . \tag{3.7.26}$$

This formula can be expressed as the Arrhenius law

$$w_{12} = \frac{V^2}{\hbar^2}\left[\frac{\pi\hbar^2}{k_B T E_r}\right]^{1/2} e^{-E_a/k_B T} \;, \tag{3.7.27}$$

where the activation energy equals

$$E_a = \frac{(J_1 - J_2 + E_r)^2}{4E_r} \;, \tag{3.7.28}$$

and E_r is the so-called reorganization energy, which provides a measure of the difference between minima of two potential wells (3.3.9) and (3.3.10)

$$E_r = \frac{1}{2}\sum_k \omega_k^2 \left(q_{k1}^0 - q_{k2}^0\right)^2 \;. \tag{3.7.29}$$

(The transition probability w_{21} is obtained by interchanging $J_1 \rightleftarrows J_2$.) The reorganization energy is expressed through function $\rho(\omega)$, (3.7.2)

$$-\int_{-a}^{a} \omega\rho(\omega)d\omega = E_r/\hbar \;. \tag{3.7.30}$$

Relation (3.7.27) is valid in the high temperature region

$$k_B T >> \hbar a \;. \tag{3.7.31}$$

It can be shown that at high temperatures (3.7.31), relation (3.7.25), obtains the form

$$| J_1 - J_2 + E_r | << \frac{k_B T}{\hbar a} E_r \;, \tag{3.7.32}$$

which may be satisfied provided the difference of the energies $|J_1\text{-}J_2|$ is not very large. Eq. (3.7.27) has been obtained by Levich [60] from his quantum theoretical treatment of the electron transfer.

Expression (3.7.27) for the transition probability depends only on the integral (3.7.30) or the sum, (2.7.29), over vibrational degrees of freedom of the Hamiltonian (3.3.14). This situation seems at first sight to be strange. Even if there is one or several vibrational degrees of freedom, one obtains a finite expression (3.7.27). As we know (1.7.22), (1.7.23), in the case of, let us say, one degree of freedom, the transition probabilitiy has a δ-function dependence on the energy of the system. The paradox is resolved if we take into account that for a very narrow energy spectrum condition (3.7.24) is not fulfilled. This means that many saddle points contribute to the sum. Thus in the case of one degree of freedom the number of saddle points that gives the same contribution tends to infinity, and expression (3.7.27) is not valid any more.

This circumstance that there are saddle points with $x_o \neq 0$, which may give a substantial contribution to the integral $I(\lambda)$, is not always taken into account. Therefore we will dwell upon this point in some detail [18]. Let us consider two models of the vibrational frequencies distribution. One of them is the Debye frequency distribution

$$f(\omega) = 3\omega^2/\omega_D^3\,; \qquad a = \omega_D\,; \qquad \int_0^{\omega_D} f(\omega)d\omega = 1\ . \tag{3.7.33}$$

In this case it can be shown that the equation for the saddle points

$$\omega_D x_o \cos\omega_D x_o = \sin\omega_D x_o\ . \tag{3.7.34}$$

In the Debye model, [18],

$$\eta^2(\omega) = E_r/\hbar\omega\ , \tag{3.7.35}$$

and in the high temperature approximation, (3.7.31)

$$\rho(\omega) = \frac{3E_r}{\hbar\omega_D}\frac{k_BT}{\hbar\omega_D}\frac{1}{\omega_D}\ . \tag{3.7.36}$$

The condition of one saddle point ($x_o = 0$) applicability, (3.7.24), takes the form

$$\frac{E_r}{\hbar\omega_D}\frac{k_BT}{\hbar\omega_D}\left(1-\frac{\sin\omega_D x_o}{\omega_D x_o}\right) >> 1, \quad x_o \neq 0\ . \tag{3.7.37}$$

It follows from (3.7.34) that at $x_o \neq 0$

$$\omega_D x_o \sim 1\ . \tag{3.7.38}$$

Therefore, at high temperatures, and the Debye model (3.7.33), the saddle point $x_o = 0$ gives the main contribution to the integral (3.7.9).

Now let us consider the Einstein model in which the vibrational frequencies are located in a narrow region near some eigenfrequency ω_o

$$f(\omega) = \begin{cases} 1/2\delta & \text{for} \quad \omega_o - \delta \le \omega \le \omega_o + \delta \\ 0 & \text{for the rest} \quad \omega > 0 \end{cases}\ . \tag{3.7.39}$$

It should be mentioned that at $\delta \to 0$, $f(\omega) \to \delta(\omega-\omega_o)$. We assume that

$$\delta \ll \omega_0 \ . \tag{3.7.40}$$

This model may be suitable for the description of optical phonons. Let us investigate saddle points $x_o \neq 0$, using equation (3.7.21). For $\delta \to 0$ it follows from equation (3.7.21) that

$$\sin \omega_0 x_0 = 0 \ . \tag{3.7.41}$$

This equation has roots

$$x_0 = \frac{2\pi k}{\omega_0} \ ; \qquad k = 0; \pm 1; \pm 2, \dots \ . \tag{3.7.42}$$

Thus, as we mentioned above, there are infinite numbers of saddle points giving the same contribution to the integral (3.7.9), and equation (3.7.27) is not valid in this case. It is clear that the Markovian equations (3.6.5) and (3.6.6) cannot be used either. The Markovian approximation cannot be satisfied since the condition (2.4.7) is not fulfilled.

Now we consider finite, but small δ, (3.7.40). In this case we assume (instead of (3.7.42)) that

$$x_0 = \frac{2\pi k}{\omega_0} + \Delta \ . \tag{3.7.43}$$

We substitute x_o (with $k = 1$) into relation (3.7.24). Taking into account that $\Delta \propto \delta^2$ [18] we obtain a relation in the lowest power of δ

$$4\pi^2 \rho(\omega_o)\delta \left(\frac{\delta}{\omega_o} \right)^2 \gg 1 \ . \tag{3.7.44}$$

According to (3.7.39) $\rho(\omega_o)\delta$ is finite when $\delta \to 0$, and is much larger than unity, (3.7.15). Thus (3.7.44) may be satisfied (i.e. only one saddle point $x_o = 0$ is taken into account) even when δ/ω_o is small. On the other hand, if (3.7.44) is not satisfied, Eq. (3.7.27) cannot be used in the Einstein model; many saddle points contribute to the integral (3.7.9).

3.8 The energy gap law

Now we examine another limiting case in which the saddle point method may be applied. This case corresponds to large energy gaps

$$|J_1 - J_2| \gg \hbar a \ ; \tag{3.8.1}$$

and a large parameter appearing in the saddle point method is

$$a|y_o| >> 1 \ . \tag{3.8.2}$$

Performing the integration by parts, one obtains from (3.7.18) and (3.7.19) equations accurate up to terms of order $(ay_o)^{-1}$

$$\lambda = \frac{e^{a|y_o|}}{a\,|\,y_o\,|}\rho\left(\frac{\lambda}{|\lambda|}a\right)a^2\cos ax_o \ , \tag{3.8.3}$$

$$\tan ax_o = \frac{ax_o}{a\,|\,y_o\,|} \ . \tag{3.8.4}$$

Solving approximately these equations and using condition (3.8.2) we get

$$a\,|\,y_o\,| = \ln\left(\frac{\lambda}{\rho\left(\frac{\lambda}{|\lambda|}a\right)a^2\cos ax_o}\right) + \ln\ln\left(\frac{|\lambda|}{\rho\left(\frac{\lambda}{|\lambda|}a\right)a^2\cos ax_o}\right) , \tag{3.8.5}$$

$$a\,|\,x_o\,| = 2\pi k + 2\pi k/|\,y_o\,|\,a \ . \tag{3.8.6}$$

We see that in this case many saddle points contribute to the integral (3.7.7) and (3.7.9). Calculating the exponent $\psi(z_o)$ and $\psi''(z_o)$ we get (neglecting terms of the order $1/a^2|y_o|^2$)

$$\psi(z_o) = -|\,\lambda y_o\,| + \frac{|\lambda|}{a} + i\frac{\lambda}{a}2\pi k + i\frac{\lambda}{a\,|\,y_o\,|\,a}\frac{2\pi k}{} \ , \tag{3.8.7}$$

$$\psi''(z_o) = -|\lambda|/a \ . \tag{3.8.8}$$

First, we will examine the contribution of the zero saddle point

$$x_o = 0; \qquad k = 0 \ , \tag{3.8.9}$$

$$a\,|\,y_o\,| = \ln\gamma + \ln\ln\gamma \ , \tag{3.8.10}$$

where

$$\gamma = \frac{|\lambda|}{\rho\left(\frac{\lambda}{|\lambda|}a\right)a^2} \ . \tag{3.8.11}$$

From Eq. (3.7.9) and Eqs. (3.8.7 – 3.8.11) we get an expression for the integral $I(\lambda)$, (3.7.7)

$$I(\lambda) = \left(\frac{2\pi}{|\lambda| a}\right)^{1/2} \exp\left\{-\frac{|\lambda|}{a}(\ln\gamma + \ln\ln\gamma - 1)\right\} . \qquad (3.8.12)$$

Using this expression, the definition of $I(\lambda)$ and expression (3.6.14), we get the following expression for the transition probability

$$w_{12} = \frac{V^2}{\hbar^2}\left(\frac{2\pi}{|\omega_{12}| a}\right)^{1/2} \exp\left[-\sum_k \eta_k^2(2\bar{n}_k + 1)\right] \exp\left\{-\frac{|\omega_{12}|}{a}\ln\left[\frac{|\omega_{12}|}{\rho(a)a^2[\bar{n}(a) + \frac{1}{2} \mp \frac{1}{2}]}\right] - 1\right\}. \qquad (3.8.13)$$

Here the "-" and "+" signs correspond to $\omega_{12} < 0$ and $\omega_{12} > 0$ respectively (for transition $1 \rightarrow 2$). When $J_1 > J_2$ ($\omega_{12} > 0$), the transition from state 1 to state 2 is possible even at zero temperature $\bar{n}(a) = 0$. The transition is performed as the multiphonon spontaneous process

$$\omega_{12} \approx Na . \qquad (3.8.14)$$

On the other hand, when $\omega_{12} < 0$; then transition from J_1 ($< J_2$) to J_2 can be performed as the multiphonon, induced by phonons, $\bar{n}(a)$, process. In this case

$$w_{12} = 0 , \quad \text{when} \quad \bar{n}(a) = 0 . \qquad (3.8.15)$$

In the expression (3.8.13) we have neglected the term $\ln\ln\gamma$ since it is much smaller than $\ln\gamma$.

For the Debye model (3.7.33)

$$\rho(a)a^2 = 3E_r/\hbar; \quad a = \omega_D , \qquad (3.8.16)$$

while for the Einstein model, (3.7.39)

$$\rho(a)a^2 = \frac{E_r}{\hbar}\frac{\omega_o}{\delta}; \quad a = \omega_o . \qquad (3.8.17)$$

Expression (3.8.13) of the energy gap law has been obtained by Englman and Jortner [41]. However, in some different form such an expression is contained in the work of Lin [53].

Now, we will clarify the validity conditions of expression (3.8.12). As we know the saddle point method may be applied provided conditions of (3.7.13) and (3.7.14) type are fulfilled. Calculating higher derivatives of $\psi(z_o)$ one can verify that conditions (3.7.13) and (3.7.14) are satisfied if

$$|\lambda|/a >> 1 \ . \tag{3.8.18}$$

This condition coincides with relation (3.8.1) which is assumed to be satisfied.

Another condition (3.8.2), which has been used in calculating the integrals, may be written, according to (3.8.5), in the form

$$\ln \frac{|\lambda|}{\rho\left(\frac{\lambda}{|\lambda|}a\right)a^2} = \ln\gamma >> 1 \ . \tag{3.8.19}$$

Before proceeding further, an important comment should be made. We have approximately calculated $a|y_o|$, (3.8.5), assuming that condition (3.8.19) is satisfied. Solving approximately Eq. (3.8.3) we get that each consecutive correction has the form $\ln\ln\gamma, \ln\ln\ln\gamma, \ldots$ All these quantities are much smaller than $\ln\gamma$. But one should remember that these quantities stand in the exponent $\psi(z_o)$, where they are multiplied by $|\lambda|/a >> 1$. Therefore the next order correction leads to the multiplication by the factor of the order of magnitude

$$\exp\left\{-\frac{|\lambda|}{a}(\ln\ln\gamma + \ln\ln\ln\gamma)\right\} \ . \tag{3.8.20}$$

Because of the assumptions (3.8.18) and (3.8.19), correction (3.8.20) essentially changes $\psi(z_o)$.

In our approximation the integral $I(\lambda)$, (3.8.12), has been calculated with the so called logarithmic accuracy. It means that the exponent index $I(\lambda)$ has been expanded into a series in which each consecutive term is much smaller than the preceding one, and only the first leading terms were preserved. But since the corrections in the exponent (their absolute values) are larger than unity, the corrections to a preexponent factor are not small. The same refers to other saddle points with $x_o \neq 0$. All of them have the same order of magnitude as $I(x_o = 0)$. But since they are calculated only with the logarithmic accuracy the summation over all saddle points z_o makes no sense. It means that the saddle point method is not appropriate for the calculation of the integral $I(\lambda)$ in the limit, (3.8.18), (3.8.19). In this case an alternative method of the integral equation [73] has been proposed.

An integral equation may be derived for the integral $I(\lambda)$, (3.7.7) and (3.7.8)

$$I(\lambda) = \int_{-\infty}^{\infty} \exp\left(\int_{-a}^{a} \rho(\omega) e^{-i\omega\tau} d\omega\right) e^{i\lambda\tau} d\tau \ , \tag{3.8.21}$$

determining transition rates (3.6.14) and (3.6.15). Performing integration by parts in the expression (3.8.21) and taking into account that in any physical situation with continuous spectrum of vibrational excitations

$$\lim_{t\to\pm\infty}\int_{-a}^{a}\rho(\omega)e^{-i\omega t}d\omega = 0 \quad , \tag{3.8.22}$$

we get

$$I(\lambda) = \frac{1}{\lambda}\int_{-a}^{a}\rho(\omega)\omega I(\lambda-\omega)d\omega \quad . \tag{3.8.23}$$

This relation constitutes the integral equation for the non-radiative transition rates as functions of the differences of energies

$$\hbar\lambda = J_1 - J_2 = \hbar\omega_{12} \quad . \tag{3.8.24}$$

According to Eqs. (3.6.14) and (3.7.1) the transition rate may be expressed through the quantity $I(\lambda)$ as follows

$$w_{12} = \frac{V^2}{\hbar^2}e^{-\int_{-a}^{a}\rho(\omega)d\omega}I(\omega_{12}) \quad . \tag{3.8.25}$$

Generally $I(\omega_{12})$ contains a singularity (the first term of the expansion of the first exponent (3.8.21) or $e^{F(\tau)}$ in (3.6.14)) at $\omega_{12} = 0$. This singularity is canceled by the second term in (3.6.14) and (3.6.15) (which equals $2\pi\delta(\omega_{21})$). In the approximations we are considering in sections 6-8, this singularity does not appear. Therefore we present the transition rate in the form (3.8.25).

A simple derivation of the functional dependence (3.7.26) on $\lambda=\omega_{12}$ may be performed directly from the integral equation (3.8.23). For this purpose one should expand $I(\lambda-\omega)$ in the integrand up to the first order terms ω

$$I(\lambda-\omega) = I(\lambda) - I'(\lambda)\omega \quad , \tag{3.8.26}$$

and to substitute for the r.h.s. of Eq. (3.8.23). As a result we obtain the differential equation

$$I'(\lambda)\int_{-a}^{a}\rho(\omega)\omega^2 d\omega + I(\lambda)\left(\lambda - \int_{-a}^{a}\rho(\omega)\omega\right) = 0 \quad . \tag{3.8.27}$$

The solution of this equation has λ-dependence (3.7.26) ($\lambda = \omega_{12}$). A condition of derivation of (3.8.27) is

$$aI'(\lambda)/I(\lambda) << 1 \quad , \tag{3.8.28}$$

which coincides with (3.7.25), or (3.7.32).

In the case of large energy gaps (3.8.1) or (3.8.18) and large parameters (3.8.19), the situation is more complicated. As a matter of fact we have failed to derive the expression for "the energy gap law" (3.8.13). The integral I(λ), (3.8.12), has the contribution of only one saddle point, $x_o = 0$. However, it has been shown that the contribution of the other saddle points $x_o \neq 0$, is not negligible. Thus expression (3.8.13) has not been proven, even with the logarithmic accuracy.

The integral equation (3.8.23) gives us the possibility to check the validity of the expressions (3.8.12) and (3.8.13) (with logarithmic accuracy). Thus we assume that

$$I(\lambda) = k \exp\left\{ -\frac{|\lambda|}{a} \ln \frac{|\lambda|}{\rho\left(\frac{\lambda}{|\lambda|} a\right) a^2} \right\} \quad . \tag{3.8.29}$$

We keep only the leading term in the exponent, (3.8.12), substitute I(λ), (3.8.29), into the integral equation (3.8.23) and assume that conditions (3.8.18) and (3.8.19) are satisfied. Then neglecting the second order terms $1/(\ln\gamma)^2$, we obtain

$$I(\lambda) = k \frac{|\lambda|}{\lambda} \frac{\exp\left\{ -\frac{|\lambda|}{a} \ln \frac{|\lambda|}{\rho(a\frac{\lambda}{|\lambda|})a^2} \right\}}{\ln\left(\frac{|\lambda|}{\rho(a\frac{\lambda}{|\lambda|})a^2} \right)} \quad . \tag{3.8.30}$$

Taking a logarithm of two parts of the Eq. (3.8.29) we get

$$\ln I(\lambda) = \left\{ -\frac{|\lambda|}{a} \ln \frac{|\lambda|}{\rho\left(a\frac{|\lambda|}{\lambda}\right) a^2} \right\} + \ln k - \ln \ln \gamma \quad , \tag{3.8.31}$$

where γ is defined by Eq. (3.8.11), and the second and third terms on the r.h.s. of Eq. (3.8.31) may be neglected. Thus the λ-dependence of I(λ) and correspondingly the energy gap law has been proven in the form (3.8.29)

$$w_{12} = k \exp\left\{ -\frac{|\lambda|}{a} \ln \frac{|\lambda|}{\rho\left(a \frac{|\lambda|}{\lambda} \right) a^2} \right\}, \quad \lambda = \omega_{12} . \tag{3.8.32}$$

For the Debye model, (3.7.33), we get

$$w_{12} = k \exp\left\{ -\frac{|J_2 - J_1|}{\hbar\omega_D} \ln \frac{|J_2 - J_1|}{3E_r\omega_D \left(n(\omega_D) + \frac{1}{2} \mp \frac{1}{2} \right)} \right\}, \tag{3.8.33}$$

and for the Einstein model, (3.7.39), we obtain

$$w_{12} = k \exp\left\{ -\frac{|J_2 - J_1|}{\hbar\omega_0} \ln \frac{|J_2 - J_1|}{E_r\omega_0 \left(n(\omega_0) + \frac{1}{2} \mp \frac{1}{2} \right)} \right\}, \tag{3.8.34}$$

where the – and + signs correspond to $J_1 < J_2$ and $J_1 > J_2$, respectively. The coefficient k remains an undetermined constant in this derivation. It is clear that these expressions mainly coincide with the energy gap law [57].

98.8 Mutual influence of the dynamic and the dissipative systems. Instabilities in thermal baths.

In the theory of relaxation processes we have two interacting systems: the dynamic system, or the system of interest and the dissipative system or thermal bath. It is conventionally assumed that during the process of the relaxation, the thermal bath preserves its state of thermal equilibrium. All the relaxation characteristics, such as relaxation times, reaction rate constants, particle free path lengths, etc., are calculated as functions of the bath temperature.

To be specific, let us consider a spin in an external magnetic field interacting with crystal lattice vibrations – a phonon bath. The question arises when and under what conditions the dynamic system (a spin in our example) may be described by the equations containing the variables (the spin components) of the subsystem only. The answer to this question depends on the relation between the relaxation times of the spin system T_{dyn} and that of the dissipative system (the phonon bath) – T_{diss}. If the relaxation of the dissipative system, e.g., due to anharmonic interactions among phonons, is much faster than that of the dynamic system (2.3.13),

$$T_{diss} << T_{dyn} \quad , \tag{3.9.1}$$

then the dissipative system very quickly reaches its equilibrium even if it was not initially in the equilibrium. In this case the equations of motion containing the spin variables only have been derived (2.3.19). The derivation is based on the smallness of the correlation function $\eta_{mn\alpha}$, (2.3.15), which in its turn depends on the fulfillment of condition (3.9.1).

The above derivation, section 2.3, is based on the tacit assumption that the spin system does not affect the rate of relaxation (or change in general) of the phonon system. Let us suppose that there is only one spin interacting with the phonon bath, which has a very large (infinite in the limit) number N of degrees of freedom. It is obvious that this spin has a negligibly small (1/N) influence on the rate of change of the phonon system. However, the model of one spin in the infinite phonon bath is by no means a realistic one. Usually there is a concentration of spins, i.e., N_{spin}/N, which is finite even when N, $N_{spin} \to \infty$. In this case it is not obvious, at least a-priori, that one can neglect the influence of spins on the rate of change of the phonon numbers. It is not obvious either that the phonon bath will be stable during the process of the relaxation, and, as we will see later on, instabilities may arise in the phonon system.

We will now consider a system of spins (or other dynamic systems). Such a system may have a continuous spectrum of energies. In our example spins may have an inhomogeneous broadening of spin frequencies. The spin of a paramagnetic atom has a definite frequency determined by a crystalline field and an applied constant field. However, the spin-spin interaction will introduce a fluctuating effective magnetic field acting upon each atom, which is superposed upon the applied field, and which has a continuous distribution of magnitudes. Let us designate the width of this frequencies distribution by δ, while the characteristic frequency width of the dissipative system is ω^*.

Now we can use the formalism presented in section 2.1. The Hamiltonian of the whole system may be presented in the form (2.1.2)

$$H = E + F + V + G = H_0 + V + G \ , \tag{3.9.2}$$

where E is the Hamiltonian of the spin system, F is the Hamiltonian of the phonon bath

$$F = \sum_k n_k \hbar\omega_k \ , \tag{3.9.3}$$

ω_k are the k-th phonon frequencies, n_k - the phonon numbers, V is the spin-phonon interaction, and G is the perturbation energy causing transitions between various phonon states (e.g., the anharmonic part of the phonon Hamiltonian (3.2.14)). We designate the total perturbation energy as

$$U = V + G \ . \tag{3.9.4}$$

Assuming fulfillment of the (2.2.22) type conditions

$$\delta >> T_{dyn}^{-1}, \ T_{diss}^{-1}; \ \omega^* >> T_{dyn}^{-1}, \ T_{diss}^{-1} \ , \tag{3.9.5}$$

we obtain equation (instead of (2.2.28))

$$\frac{\partial \rho_{n\alpha;n\alpha}}{\partial t} = -\sum_{\beta,m} w_{n\alpha;m\beta}\left(\rho_{n\alpha;n\alpha} - \rho_{m\beta;m\beta}\right) . \tag{3.9.6}$$

Here $\rho_{n\alpha;n\alpha}$ is the probability of the state (α,n), α - is the phonon state, n – is the spin system state, and

$$w_{n\alpha;m\beta} = \frac{2\pi}{\hbar}\left|U_{n\alpha;m\beta}\right|^2 \delta\left(E_n + F_\alpha - E_m - F_\beta\right) . \tag{3.9.7}$$

Assuming that the diagonal in spin indices m,n matrix elements are

$$V_{m\alpha;m\beta} = 0 , \tag{3.9.8}$$

we obtain instead of (3.9.6) the following equation

$$\frac{\partial \rho_{n\alpha;n\alpha}}{\partial t} = -\frac{2\pi}{\hbar}\sum_{m\beta}\left|V_{n\alpha;m\beta}\right|^2 \delta\left(E_n + F_\alpha - E_m - F_\beta\right)\left(\rho_{n\alpha;n\alpha} - \rho_{m\beta;m\beta}\right)$$

$$-\sum_{\beta} w_{\alpha\beta}\left(\rho_{n\alpha;n\alpha} - \rho_{n\beta;n\beta}\right) , \tag{3.9.9}$$

where $w_{\alpha\beta}$ is determined by Eq. (2.2.27).

We will look for solutions of Eq. (3.9.9) in the form

$$\rho_{n\alpha;n\alpha} = \sigma_n(t)P_\alpha(t); \qquad \sigma_n \equiv \sigma_{nn}; \qquad P_\alpha \equiv \rho_{\beta\beta} . \tag{3.9.10}$$

Substituting these relations in the r.h.s. of Eq. (3.9.9), performing the summation over α and then over n we get

$$\frac{d\sigma_n}{dt} = -\sum_{m,\alpha,\beta} W_{n\alpha;m\beta}\left(P_\alpha\sigma_n - P_\beta\sigma_m\right) , \tag{3.9.11}$$

$$\frac{dP_\alpha}{dt} = -\sum_{n,m,\beta} W_{n\alpha;m\beta}\left(P_\alpha\sigma_n - P_\beta\sigma_m\right) - \sum_{\beta} w_{\alpha\beta}\left(P_\alpha - P_\beta\right) , \tag{3.9.12}$$

where

$$W_{n\alpha;m\beta} = \frac{2\pi}{\hbar}\left|V_{n\alpha;m\beta}\right|^2 \delta\left(E_n + F_\alpha - E_m - F_\beta\right) , \tag{3.9.13}$$

and

$$w_{\alpha\beta} = \frac{2\pi}{\hbar}|G_{\alpha\beta}|^2 \delta(F_\alpha - F_\beta) \quad . \tag{3.9.14}$$

Equations (3.9.11) and (3.9.12) are equations of motion of interacting spin-phonon systems. As opposed to the system with one spin, the time development of the phonon subsystem probabilities distributions P_α is determined not only by $w_{\alpha\beta}$ (anharmonic interactions between phonons), but also by the interaction with the spin subsystem.

A certain remark has to be made about the validity of Eqs. (3.9.11) and (3.9.12). A representation of $\rho_{n\alpha;n\alpha}$ as a product of density matrices of the spin and phonon subsystems is a kind of the "mean-field" assumption. Usually such an assumption is justified if one can neglect corresponding fluctuations (see e.g. [74]). In our case it means that the correlations $\eta_{mn\alpha}$, (2.3.15), are neglected. As it has been in the near-equilibrium case, $P_\alpha \approx P_\alpha^{eq}$, the correlations $\eta_{mn\alpha}$ can be neglected provided condition (2.3.13) of the fast relaxation of the dissipative system is satisfied. One may expect that "the mean-field approximation", (3.9.11, 3.9.12), is valid for time-dependent $P_\alpha(t)$, if condition (2.3.13) is satisfied. It has to be mentioned that (3.9.11) and (3.9.12) type equations (without the last sum in the r.h.s. of (3.9.12)), were suggested by Lax [75] in 1966.

From the master equations (3.9.11) and (3.9.12) one can derive [76] relatively simple equations for the mean values of phonon numbers

$$\bar{n}_q = \sum_{n_1, n_2, \ldots, n_q, \ldots} P(n_1, n_2, \ldots, n_q, \ldots) n_q \quad , \tag{3.9.15}$$

and for the mean values of the differences of the populations n_j between ground and excited states of the j-th spin

$$2\bar{r}_{j3} = \bar{n}_j^+ - \bar{n}_j^+, \qquad \bar{n}_j^+ + \bar{n}_j^- = 1 \quad , \tag{3.9.16}$$

$$\bar{r}_{j3} = \sum_{\ldots, r_{j3}, \ldots} \sigma(\ldots, r_{j3}, \ldots) r_{j3} \quad , \tag{3.9.17}$$

where $P(\ldots,n_q,\ldots)$ and $\sigma(\ldots,r_{j3},\ldots)$ are explicit expressions for P_α and σ_n in Eqs. (3.9.11) and (3.9.12).

We consider a specific and relatively simple model [76]. This model is supposed to describe an ensemble of two-level systems embedded in the phonon bath. The differences of the energies between upper and lower levels $\hbar\omega_j$ are assumed to be quasicontinuously distributed over a certain energy band with a characteristic bandwidth $\hbar\delta$. The phonon frequencies are distributed over the phonon zone with a characteristic bandwidth $\hbar\omega^*$. The model Hamiltonian of such a system may be chosen as

$$H=\sum_j r_{j3}\hbar\omega_j+\sum_q \hbar\omega_q a_q^+ a_q+\sum_{j,q} V_{jq}\left(r_j^+ a_q+r_j^- a_q^+\right)+G \ . \qquad (3.9.18)$$

Here a_q^+, a_q are the creation and annihilation operators of phonons with energies $\hbar\omega_q$; r_3, $r^{\pm}=r_1 \pm ir_2$ are the effective spin operators (3.3.1), (3.3.2) and (3.3.7). In the case of real spins they coincide with the projections of spin $\frac{1}{2}$ operators on the axis x, y, z.

Equations for the mean values $\bar{n}_q$ and $\bar{n}_j^{\pm}$ take the form

$$\frac{d\bar{n}_j^+}{dt}=-\frac{d\bar{n}_j^-}{dt}=\frac{2\pi}{\hbar^2}\sum_q |V_{jq}|^2\left[(\bar{n}_q+1)\bar{n}_j^+-n_q\bar{n}_j^-\right]\delta(\omega_j-\omega_q) \ ; \qquad (3.9.19)$$

$$\frac{d\bar{n}_q}{dt}=\frac{2\pi}{\hbar^2}\sum_j |V_{jq}|^2\left[(\bar{n}_j^+-n_j^-)\bar{n}_q+\bar{n}_j^+\right]\delta(\omega_j-\omega_q)-W_q(\bar{n}_q-n_q^{\circ}) \ , \qquad (3.9.20)$$

where W_q is the relaxation rate towards the thermal equilibrium (this quantity W_q is the phenomenological constant, reflecting the relaxation described by the last term in the r.h.s. of Eq. (3.9.12)), and

$$n_q^{\circ}=\left[\exp(\beta\hbar\omega_q)-1\right]^{-1}; \qquad \beta=(k_B T)^{-1} \ . \qquad (3.9.21)$$

(The simplified form of the relaxation term W_q, in the r.h.s. of (3.9.20), does not affect the forthcoming analysis.) As has been mentioned above (in the beginning of the section), in the case of a finite concentration of spins (as opposed to one spin), these spins can give a finite contribution to the relaxation rates of phonons. This contribution is presented by the first term in the r.h.s. of Eq. (3.9.20). For one spin the contribution of this term is infinitesimally small.

The conventional assumption that spins do not influence the phonon relaxation is valid provided

$$\sigma=\frac{2\pi}{\hbar^2 W_q}\left|\sum_j |V_{jq}|^2(\bar{n}_j^+-n_j^-)\delta(\omega_j-\omega_q)\right| \ll 1 \ . \qquad (3.9.22)$$

On the other hand, if factor

$$\sigma \sim 1 \ , \qquad (3.9.23)$$

one cannot neglect the influence of the spin system on the rate of change of phonon numbers. In this case the joint system of equations (3.9.19) and (3.9.20) determine relaxation of both $\bar{n}_j^{\pm}$ and n_q.

The spin system may also cause the instability in the phonon system. Let us assume that there exists a small perturbation in some mode $\bar{n}_q$ of the phonon bath. Under certain conditions this small perturbation $\bar{n}_q$ of the phonon bath begins to rise exponentially. It is easy to see from Eq. (3.9.20) that the condition for the instability has the form

$$\frac{2\pi}{\hbar^2 W_q}\sum_j |A_{jq}|^2 (\bar{n}_j^+ - \bar{n}_j^-)\delta(\omega_j - \omega_q) > 1 \quad . \tag{3.9.24}$$

We see that the instabilities in the phonon bath arise provided there is inversion of the spin population

$$\bar{n}_j^+ > \bar{n}_j^- \quad . \tag{3.9.25}$$

(In the thermal equilibrium $\bar{n}_j^- > \bar{n}_j^+$.) Of course, condition (3.9.24) depends on time through $\bar{n}_j^+(t) - \bar{n}_j^-(t)$. The instability exists only at time intervals satisfying (3.9.24). On the other hand, if $\bar{n}_{j'}^+ - \bar{n}_j^-$ is sustained by some external source (like a maser), then $\bar{n}_q$ increases exponentially until the steady state is reached. This steady state may be similar to that achieved in quantum oscillators – it may be determined by the anharmonicity of the bath vibrations.

Let us analyze condition (3.9.24) (or, more generally, condition (3.9.23)). For this purpose we introduce parameters

$$\gamma_o = \frac{2\pi}{\hbar^2}\sum_q |V_{jq}|^2 \delta(\omega_j - \omega_q) = \frac{2\pi}{\hbar^2}|\bar{V}_{jq}|^2 \rho(\omega_j) \quad , \tag{3.9.26}$$

which is the spin relaxation rate, and

$$\Gamma_o = \frac{2\pi}{\hbar^2}\sum_j |V_{jq}|^2 \delta(\omega_j - \omega_q) = \frac{2\pi}{\hbar^2}|\tilde{V}_{jq}|^2 f(\omega_q) \quad , \tag{3.9.27}$$

which is the rate of the phonon relaxation due to the interaction with spins.

We assume that the mean values

$$|\bar{V}_{jq}|^2 \approx |\tilde{V}_{jq}|^2 \quad , \tag{3.9.28}$$

and $\rho(\omega_j)$, $f(\omega)$ are the frequency distributions of phonon and spin frequencies, respectively

$$\int \rho(\omega_q)d\omega_q = N \quad , \tag{3.9.29}$$

$$\int f(\omega_j)d\omega_j = \tilde{N} \quad . \tag{3.9.30}$$

Here N and $\tilde{N}$ are numbers of phonon and spin degrees of freedom, respectively. Using these designations we can rewrite the condition (3.9.24) in the form

$$\left(\overline{n}_j^+ - \overline{n}_j^-\right)\frac{f(\omega_q)}{\rho(\omega_j)}\gamma_o > W_q \quad , \tag{3.9.31}$$

where $\overline{n}_j^+ - \overline{n}_j^-$ is a certain mean value in the sum (3.9.24). This condition has a quite obvious interpretation. The expression $(\overline{n}_j^+ - \overline{n}_j^-)f(\omega_q)$ is the effective number of excited spins per unit frequency, $(\overline{n}_j^+ - \overline{n}_j^-)f(\omega_q)\gamma_o$ is the rate of energy input in the mode n_q. Correspondingly, $\rho(\omega_j)W_q$ is the rate of output of the energy from the mode q (with $\omega_j \approx \omega_q$) to all other modes via the mechanism described phenomenologically by the rate constant W_q. Thus, condition (3.9.24) means that input of energy is larger than output.

We assume the Debye model for the phonon frequency distribution

$$\rho(\omega) = 3N\frac{\omega^2}{\omega_D^3} \quad , \tag{3.9.32}$$

and the Einstein model for the spin frequency distribution

$$f(\omega) = \tilde{N}\begin{cases} \dfrac{1}{\delta} \quad \text{when} \quad \omega_0 - \dfrac{\delta}{2} \le \omega \le \omega_0 + \dfrac{\delta}{2} \\ 0 \quad \text{for} \quad \omega < \omega_0 - \dfrac{\delta}{2} \quad \text{and} \quad \omega > \omega_0 + \dfrac{\delta}{2}. \end{cases} \tag{3.9.33}$$

With these models condition (3.9.31) obtains the form

$$\sigma = \frac{N^+ - N^-}{N}\frac{\gamma_0\omega_D^3}{3\omega_0^2\delta \cdot W_q} > 1 \quad . \tag{3.9.34}$$

First the phenomenon of a nonequilibrium thermal bath was explored by Van Vleck [77], 60 years ago: "It is the purpose of the present article to point out that it is impossible for the lattice oscillators to preserve a constant temperature in the face of frequent energy transfers with the spin". If the spin-phonon coupling is strong, and phonons are generated in the narrow "bottleneck" proportional to $\omega_0^2\delta$ (factor $\sigma > 1$), then the thermalization does not happen. The increase in the number of the phonons was named "avalanche". This phenomenon happens at helium temperatures, when so-called direct processes take place. At liquid-air temperatures the interaction between spin and lattice is secured by means of second-order Raman-like processes;

when a spin scatters a lattice quantum of comparatively high energy, the bottleneck is "destroyed".

Phonon avalanche and bottleneck effects have been the subject of various studies [70,78-84]. We have considered here the inhomogeneous broadening of spin frequencies. The case of homogeneous broadening characterized by T_2^{-1} has been considered in Ref. 76. Apart from nonessential numerical factors, parameter σ (3.9.34), can be obtained by replacing δ by T_2^{-1}.

CHAPTER IV

INTERACTION WITH PHOTONS

Interaction of matter with the electromagnetic field is one of the important subjects of the scientific research. When eigenfrequencies of the electromagnetic field form the continuum, interaction with the electromagnetic field leads to the irreversible processes. These irreversible processes are, to a certain extent, similar to those considered in the preceding chapter (interaction with phonons). Nevertheless, the irreversible processes due to the interaction with photons have their own specific features. Thus the relaxation of a system of two-level molecules may have a collective character. Therefore, a substantial part of this chapter is devoted to the phenomenon of the superradiance. The notions of spontaneous and stimulated emission traditionally belong to the region of the matter-field interaction. However, as will be shown, these notions (spontaneous and stimulated emissions) characterize the interaction with phonons as well.

4.1 Interaction of matter with the electromagnetic field

In quantum theory the Hamiltonian of a system of particles interacting with an electromagnetic field can be presented as

$$H = \frac{1}{2}\sum_{v}\left(p_v^2 + \omega_v^2 q_v^2\right) + \sum_{k}\frac{1}{2m_k}\left(\vec{p}_k - \frac{e_k}{c}\vec{A}(k)\right)^2 + \frac{1}{2}\sum_{i\neq k}\frac{e_i e_k}{r_{ik}}, \qquad (4.1.1)$$

where m_k, $\vec{p}_k$ are the mass and canonical momentum of the k-th particle respectively; r_{ik} is the distance between the i-th and k-th particles with charges e_i and e_k, $\vec{A}(k)$ is the vector-potential at the location of the k-th particle

$$\vec{A}(k) = \sum_{v} q_v \vec{A}_v(k)\,; \quad \mathrm{div}\vec{A} = 0\ . \qquad (4.1.2)$$

The variables p_v and q_v are canonical variables describing the electromagnetic field

$$[q_v, p_{v'}] = i\hbar\delta_{vv'}\ . \qquad (4.1.3)$$

To present the Hamiltonian (4.1.1) in the standard form (2.1.2) appropriate for the employment of the theory of irreversible processes, it is convenient to extract from the Hamiltonian (4.1.1) the energy of the interaction with the radiation field

$$H = H_0 + V = E + F + V\ , \qquad (4.1.4)$$

where H_0 is the unperturbed Hamiltonian of the system consisting of particles interacting with each other in accordance with Coulomb's law

$$E = \sum_k \frac{p_k^2}{2m_\alpha} + \frac{1}{2}\sum_{i \neq k} \frac{e_i e_k}{r_{ik}} , \tag{4.1.5}$$

plus the free radiation field

$$F = \frac{1}{2}\sum_\nu \left(p_\nu^2 + \omega_\nu^2 q_\nu^2\right) , \tag{4.1.6}$$

whilst the energy of the interaction with the field is

$$V = -\sum_k \frac{e_k}{m_k c}\left(\vec{p}_k \cdot \vec{A}(k)\right) + \sum_k \frac{e_k^2}{2m_k c^2}\vec{A}^2(k) . \tag{4.1.7}$$

In deriving Eq. (4.1.7) the Coulomb gauge, divA = 0, has been assumed. In the solution of a number of problems (in particular those connected with spontaneous and stimulated emission of radiation) we can neglect the second term in the right-hand side of Eq. (4.1.7). We can rewrite the first term in the form

$$V = -\sum_\nu B_\nu q_\nu , \tag{4.1.8}$$

where B_ν is an operator acting only on the particle variables

$$B_\nu = \sum_k \frac{e_k}{m_k c}\left(\vec{p}_k \cdot \vec{A}_\nu(k)\right) . \tag{4.1.9}$$

Let us now examine the case when the dimensions of the system of particles are small compared with the wavelength of the radiation (or compared with characteristic dimensions of the inhomogeneity of the vector-potential $\vec{A}(\vec{r}_k)$). This is generally the case for atoms and molecules over a very wide range of frequencies, including the optical range. This means that the size of the atom or molecule is much smaller than the wavelength. In this case the interaction energy with the electromagnetic field can be written in the form

$$V = \frac{1}{c}\left(\vec{A}\dot{\vec{d}}\right) - \left(\vec{\mu} \cdot \vec{H}\right) , \tag{4.1.10}$$

where

$$\vec{d} = \sum_k e_k \vec{r} , \tag{4.1.11}$$

and $\vec{\mu}$ is the magnetic moment of the molecule (taking into account both orbital motion and the spin of the system). In the equation (4.1.10) the difference between the kinetic and canonical momenta has been neglected (as well as other terms of the order of e_k^2).

4.2 A system of two-level molecules interacting with the electromagnetic field

We consider the case when only two levels of the molecule (or atom) are relevant, and the influence of the other levels may be neglected. In this case the dipole moments of the molecule - electric $\vec{d}$ and magnetic $\vec{\mu}$ can be expressed through the effective spin operators r_1, r_2 and r_3 (see section 3.3)

$$\frac{1}{c}\dot{\vec{d}} = \vec{e}_1 r_1 + \vec{e}_2 r_2 \quad , \tag{4.2.1}$$

$$\vec{\mu} = \vec{m}_1 r_1 + \vec{m}_2 r_2 + \vec{m}_3 r_3 \quad , \tag{4.2.2}$$

where constants $\vec{e}_1, \vec{e}_2, \vec{m}_1, \vec{m}_2$, and $\vec{m}_3$ are determined by the molecule's electric and magnetic dipoles matrix elements. Operators of the vector-potential $\vec{A}$ and magnetic field $\vec{H}$ can be expressed through the normal modes functions $\vec{A}_v$

$$\vec{A}(\vec{x}) = \sum_v q_v \vec{A}(\vec{x}); \quad \vec{H}(\vec{x}) = -\sum_v \omega_v \vec{H}_v(\vec{x}); \quad \omega_v \vec{H}_v = \mathrm{rot}\vec{A}_v \ . \tag{4.2.3}$$

The Hamiltonian (4.1.1) of the system of two-level molecules may be written in the form

$$H = \frac{1}{2}\sum_v \left(p_v^2 + \omega_v^2 q_v^2\right) - \sum_j \hbar(\vec{r}_j \cdot \vec{K}(j)) \quad , \tag{4.2.4}$$

where

$$\vec{K}(j) = \left\{ \frac{(\vec{A}(j)\cdot\vec{e}_1)}{\hbar} + \frac{(\vec{H}(j)\cdot\vec{m}_1)}{\hbar}; \frac{(\vec{A}(j)\cdot\vec{e}_2)}{\hbar} + \frac{(\vec{H}(j)\cdot\vec{m}_2)}{\hbar}; -\omega_j + \frac{(\vec{H}(j)\vec{m}_3)}{\hbar} \right\} \ . \tag{4.2.5}$$

The alternative form of the Hamiltonian (4.2.4) is

$$H = \frac{1}{2}\sum_v \left(p_v^2 + \omega_v^2 q_v^2\right) + \sum_j r_{3j}\hbar\omega_j - \sum_j \hbar\left(\vec{r}_j \cdot \vec{\tilde{K}}(j)\right) \ , \tag{4.2.6}$$

where $\vec{\vec{K}}$ differs from $\vec{K}$, (4.2.5) by absence of $-\omega_j$ in the 3-component of $\vec{R}$. It should be noticed that we have assumed the dipole approximation for each molecule, (4.1.10), while the whole system is not limited by the dipole approximation .

The equation of motion for the effective spin $\vec{r}_j$ and the field equation of motion follow from the Hamiltonian (4.2.4)

$$\dot{\vec{r}}_j = \vec{r}_j x \vec{K}(j) \quad , \tag{4.2.7}$$

$$\ddot{q}_v + \omega_v^2 q_v = \sum_j \left[\vec{A}_j(j) \cdot (\vec{e}_1 r_{1j} + \vec{e}_2 r_{2j}) - \omega_v \vec{H}_v(j) \cdot (\vec{m}_1 r_{1j} + \vec{m}_2 r_{2j} + \vec{m}_3 r_{3j}) \right] \ . \tag{4.2.8}$$

These are equations for operators $\vec{r}$ and q_v. We define the effective spin distribution by the relation

$$\vec{s}(\vec{x}) = \sum_j \vec{r}_j \delta(\vec{x}_j - \vec{x}) \ . \tag{4.2.9}$$

The total spin of the system is then

$$\vec{R} = \int \vec{s}(\vec{x}) dV = \sum_j \vec{r}_j \ . \tag{4.2.10}$$

The equations of motion for the quantities $\vec{s}$ and q_v become

$$\dot{\vec{s}} = \vec{s} x \vec{K} \quad , \tag{4.2.11}$$

$$\ddot{q}_v + \omega_v^2 q_v = \int \vec{A}_v(\vec{x}) \cdot [\vec{e}_1 s_1 + \vec{e}_2 s_2] dV - \omega_v \int \vec{H}_v(\vec{x}) \cdot [\vec{m}_1 s_1 + \vec{m}_2 s_2 + \vec{m}_3 s_3] dV \ . \tag{4.2.12}$$

Quantum equations of motion (4.2.7), (4.2.8), (4.2.11) and (4.2.12) take into account interactions between the spins and the electromagnetic field only, (4.2.6). They do not take into account the relaxation of the electromagnetic field and spins due to other interactions (spin-lattice interactions, photon-phonon interactions, etc.). These interactions are described by the term G in the Hamiltonian (2.1.2). The latter was not taken into account in the Hamiltonian (4.1.4 – 4.1.7).

Equations (4.2.7) and (4.2.11) are analogues of the Bloch equations for real spins. The relevance of the Bloch equations for the description of the maser was first recognized by Feynmann, Vernon and Helwarth [85]. Taking into account (phenomenologically) the above mentioned interactions, we can write equations for the mean values of the effective spins and field (assuming that all spins are equivalent $\omega_j = \omega_0$)

$$\dot{s}_1 + \omega_0 s_2 + \frac{1}{T_2} s_1 + \frac{1}{\hbar} s_3 \left[(\vec{A} \cdot \vec{e}_2) + (\vec{H} \cdot \vec{m}_2) \right] - s_2 \frac{(\vec{H} \cdot \vec{m}_3)}{\hbar} = 0 \quad , \tag{4.2.13}$$

$$\dot{s}_2 - \omega_o s_1 + \frac{1}{T_2} s_2 - \frac{1}{\hbar} s_3 \left[(\vec{A} \cdot \vec{e}_1) + (\vec{H} \cdot \vec{m}_1) \right] + s_1 \frac{(\vec{H} \cdot \vec{m}_3)}{\hbar} = 0 \quad , \qquad (4.2.14)$$

$$\dot{s}_3 = -\frac{1}{T_1}\left(s_3 - s_3^0\right) - \frac{1}{\hbar}\left(\vec{A} \cdot [\vec{e}_1 s_2 - \vec{e}_2 s_1]\right) - \frac{1}{\hbar}\left(\vec{H} \cdot [\vec{m}_1 s_2 - \vec{m}_2 s_1]\right) , \qquad (4.2.15)$$

$$\ddot{q}_v + \gamma_v \dot{q}_v + \omega_v^2 q_v = \int \left(\vec{A}_v(\vec{x}) \cdot [\vec{e}_1 s_1 + \vec{e}_2 s_2] \right) dV$$

$$-\omega_v \int \vec{H}_v(\vec{x}) \cdot \left([\vec{m}_1 s_1 + \vec{m}_2 s_2 + \vec{m}_3 s_3] \right) dV \quad , \qquad (4.2.16)$$

where $\vec{A}$, $\vec{H}$, $\vec{A}_v$ and $\vec{H}_v$ are defined by Eq. (4.2.3). Here T_1 and T_2 and γ_v are phenomenologically introduced relaxation constants. They take into account the term G. Equations of the (4.2.13) – (4.2.16) type are called Bloch-Maxwell equations. They were derived by Fain [51] in 1959 (see also Ref. 87).

4.3 Quantum theory of spontaneous and stimulated emission in a system of two-level molecules

In this section we shall discuss spontaneous and stimulated emission from a collection of two-level molecules that do not interact with each other in the absence of the field. The two-level molecules are described by the effective spin operators r_1, r_2 and r_3, We shall further suppose that all molecules have the same energy level difference $\hbar\omega_o$. In the case when the dimensions of the system of molecules is much less than a wavelength, i.e. the dipole approximation for the whole system, the Hamiltonian of the system can be written in the form (see (4.2.6))

$$H = \hbar\omega_o R_3 + \frac{1}{2}\sum_{\gamma}\left(p_v^2 + \omega_v^2 q_v^2\right) - \left(\vec{A}\vec{e}_1\right) R_1 \quad , \qquad (4.3.1)$$

where the operator of the whole spin

$$R_1 = \sum_j r_{1j}; \qquad R_2 = \sum_j r_{2j}; \qquad R_3 = \sum_j r_{3j} \quad . \qquad (4.3.2)$$

It is worthwhile mentioning that $2R_3$ has meaning of population difference of upper and lower states

$$2R_3 = N_+ - N_- , \quad N_+ + N_- = N \quad , \qquad (4.3.3)$$

while R_1 and R_2 describe the dipole moment of the system. Several assumptions have been made here. We put $\vec{m}_1 = \vec{m}_2 = \vec{m}_3 = 0$, which means neglecting the magnetic dipole system. These terms are small in comparison with the electric dipole terms $\vec{e}_1 \neq 0$. We also put $\vec{e}_2 = 0$, just for the sake of simplicity. This assumption does not affect any further conclusions.

It is worthwhile to consider in some detail the validity of the dipole approximation for the whole system. The third term in Eq. (4.2.6) can be written in the form (see also (4.2.3))

$$\hat{V} = -\sum_{j,\nu} q_\nu \left(\vec{A}_\nu(j)\cdot\vec{e}_1\right) r_{1j} \quad . \tag{4.3.4}$$

In the general case the transition to the dipole approximation (4.3.1) is possible if the dimensions of the system are smaller than <u>all</u> wavelengths (or other space characteristics of $\vec{A}_\nu(\vec{x})$). However, such a transition is impossible, since the sum (4.3.4) always contains modes $\vec{A}_\nu$ with wavelengths which are smaller than the dimensions of the system. The usual assumption is that the dominant contribution to the sum (4.3.4) is connected with modes ν having frequencies $\omega_\nu \approx \omega_o$. The induced and spontaneous radiation takes place at these frequencies. However, below we will see that there are situations in which one cannot neglect the contribution of other modes, $\omega_\nu \neq \omega_o$.

Now we will analyze the Hamiltonian (4.3.1). (We will neglect for a while the contribution of higher frequency modes.) The quantum equations of motion following from the Hamiltonian (4.3.1) are of the form

$$\dot{p}_\nu = \omega_\nu^2 q_\nu + (\vec{e}_1\cdot\vec{A}_1)R_1, \qquad \dot{q}_\nu = p_\nu \ ,$$
$$\dot{R}_1 = -\omega_o R_2, \quad \dot{R}_2 = \omega_o R_1 + \frac{(\vec{A}\cdot\vec{e}_1)}{\hbar}R_3; \quad \dot{R}_3 = -\frac{1}{\hbar}(\vec{A}\cdot\vec{e}_1)R_2 \quad . \tag{4.3.5}$$

Let us find the change with the time of the ν-th mode energy operator

$$H_\nu = \frac{1}{2}\left(p_\nu^2 + \omega_\nu^2 q_\nu^2\right) \quad . \tag{4.3.6}$$

The derivative of this quantity can be found in the usual way

$$\frac{dH_\nu}{dt} = (\vec{e}_1\cdot\vec{A}_\nu)R_1 p_\nu \quad . \tag{4.3.7}$$

Including terms up to the second order of smallness over small parameter $(\vec{e}_\nu\vec{A}_\nu)$ we find

$$\frac{dH_\nu}{dt} = (\vec{e}_1 \cdot \vec{A}_\nu) R_1^{(0)} p_\nu^{(0)} + (\vec{e}_1 \cdot \vec{A}_\nu) R_1^{(1)} p_\nu^{(0)} + (\vec{e}_1 \cdot \vec{A}_\nu) R_1^{(0)} p_\nu^{(1)} \quad . \tag{4.3.8}$$

Here $R_1^{(0)}, p_\nu^{(0)}$, and $R_1^{(1)}, p_\nu^{(1)}$ are the zero and first order quantities respectively. The mean value (4.3.8) gives the intensity of radiation of the ν-th mode.

Let us first consider the mean value of the first term in the r.h.s. of (4.3.8)

$$\left(\frac{d\bar{H}_\nu}{dt}\right)_1 = (\vec{e}_1 \cdot \vec{A}_\nu) \bar{R}_1^{(0)} \bar{p}_\nu^{(0)} \quad . \tag{4.3.9}$$

The mean values $\bar{R}_1^{(0)}$ and $\bar{p}_\nu^{(0)}$ have the form

$$\bar{R}_1^{(0)} = \bar{R}_1(0)\cos\omega_o t - \bar{R}_2(0)\sin\omega_o t \equiv R\sin(\omega_o t + \varphi) \quad , \tag{4.3.10}$$

$$\bar{p}_\nu^{(0)}(t) = \bar{p}_\nu(0)\cos\omega_\nu t - \omega_\nu \bar{q}_\nu(0)\sin\omega_\nu t \equiv p\sin(\omega_\nu t + \psi) \quad . \tag{4.3.11}$$

The induced emission and absorption are determined by the sign of $d\bar{H}_\nu / dt$; when it is positive we have the induced emission, the negative sign corresponds to the absorption. If, for the sake of simplicity we consider that $\omega_0 = \omega_\nu$, then the time-averaged value ($t \gg 2\pi/\omega_o$) of (4.3.9) is

$$\left(\frac{d\bar{H}_\nu}{dt}\right)_1 = \frac{Rp}{2}\cos(\varphi - \psi) \quad . \tag{4.3.12}$$

Thus the sign of the radiation intensity depends on the phase relation between the field and the system. This phase dependent stimulated emission (absorption) may occur both in the classical and in the quantum theory. Moreover, this is not what is meant by stimulated emission (absorption) in the textbooks; the latter is connected with the phase-independent part of (4.3.8): the second and third terms in the r.h.s. of (4.3.8). They are much smaller than $(dH_\nu/dt)_1$, since they are proportional to the second order of a small parameter, proportional to $(\vec{e}_1 \cdot \vec{A}_\nu)$. However, if

$$\bar{p}_\nu(0) = 0; \qquad \bar{q}_\nu(0) = 0; \qquad \text{or} \qquad \bar{R}_1(0) = \bar{R}_2(0) = 0; \tag{4.3.13}$$

(or both of them are zero), then the first order term (4.3.9) vanishes.

We now assume that conditions (4.3.13) (or one of them) are fulfilled. Then, as a result of averaging, only the second and third terms in the r.h.s. of Eq. (4.3.8) survive. These terms can be presented in the form [88,21]

$$\left(\frac{d\bar{H}_\nu}{dt}\right)_2 = \pi\frac{(\vec{e}_1\vec{A}_\nu)}{4}\left[\overline{R_1^2}(0) + \overline{R_2^2}(0) + \frac{\bar{R}_3}{\hbar\omega_\nu}\left(\overline{p_\nu^2} + \omega_\nu^2\overline{q_\nu^2}\right)\right]\delta(\omega_0 - \omega_1) \quad . \tag{4.3.14}$$

For the system of N two-levels molecules located in the volume $L^3 << \lambda^3$ (λ is the wavelength of the emitted radiation) expression (4.3.14) can describe the superradiance. We will consider this phenomenon later on, in sections (4.5 - 4.8).

4.4 Spontaneous emission vs. vacuum fluctuations

The question about the relation between spontaneous emission and vacuum fluctuations has a long history. Enough to mention that as early as in 1935 Weisskopf [89] claimed that spontaneous emission is ascribed entirely to the zero-point fluctuations, or vacuum fluctuations, of the electromagnetic field. In 1939 Ginzburg [90] showed that spontaneous emission is not purely a quantum phenomenon (as it has to be if it is due to the vacuum fluctuations) and exists in classical theory as well. The radiation damping of the classical oscillator is the example of classical spontaneous emission.

The role of vacuum fluctuations has been clarified by Fain [88,21,91]. It has also been treated by Dalibard et al [92]. We start from Eq. (4.3.14) and consider the case of a one two-level molecule. In this case, $r_1^2 = r_2^2 = \frac{1}{4}$, (3.3.2); $r_3 = \frac{1}{2}(n_+ - n_-)$, (4.3.3),

$$I_\nu = I_\nu^0 \left[\frac{1}{2} + \frac{1}{2}(n_+ - n_-) \frac{\overline{p_\nu^2} + \omega_\nu^2 \overline{q_\nu^2}}{\hbar\omega_\nu} \right] , \tag{4.4.1}$$

where n_+ and n_- are the populations of the upper and lower levels of the two-level molecule, $n_+ + n_- = 1$; I_ν is the intensity of the emission of the two-level molecule, for arbitrary n_+ and n_-, while I_ν^0 is the intensity of spontaneous emission of the molecule, from the initial state $n_+ = 1$, $n_- = 0$.

Consider only the term connected with spontaneous emission in expression (4.4.1), which can be evaluated if the radiation field was initially in a vacuum state. This does not imply that the energy of the radiation field

$$\bar{H}_\nu(0) = \frac{1}{2}\left[\overline{p_\nu^2(0)} + \omega_\nu^2 \overline{q_\nu^2(0)}\right] = 0 \ . \tag{4.4.2}$$

In fact, the mean energy of the radiation field in the ν-th mode is of the form

$$\bar{H}_\nu(0) = \left(\bar{n}_\nu + \frac{1}{2} \right) \hbar\omega_\nu \ , \tag{4.4.3}$$

where $\bar{n}_\nu$ is the mean number of photons in the ν-th mode. The term $\frac{1}{2}\hbar\omega_\nu$ is the so-called zero-point energy of the field. For a given choice of the Hamiltonian, (4.3.6), the zero-point energy serves as a measure of the zero-point fluctuations of the field, i.e.

of the quantity $\frac{1}{2}\,(\overline{p_\nu^2}+\omega_\nu^2\overline{q_\nu^2})$. Generally speaking, however, the zero-point energy does not describe vacuum fluctuations. In fact, we can carry out a transformation of the Hamiltonian of the field

$$H'_\nu = H_\nu - \frac{1}{2}\hbar\omega_\nu \, . \tag{4.4.4}$$

As a result of this transformation the equations remain unchanged (therefore $\frac{1}{2}\,(p_\nu^2+\omega_\nu^2 q_\nu^2)$ also remains unchanged) and the zero-point energy disappears: $\overline{H'_\nu} = 0$.

Now we can evaluate the role of vacuum fluctuations

$$n_\nu = 0 \quad \text{and} \quad \frac{1}{2}(\overline{p_\nu^2}+\omega_\nu^2\overline{q_\nu^2}) = \frac{1}{2}\hbar\omega_\nu \; , \tag{4.4.5}$$

in the spontaneous radiation. The intensity of the spontaneous emission of the two-level molecule is:

$$I = I_0\left(\frac{1}{2}+\frac{1}{2}(n_+ - n_-)\right) = I_0\left(\frac{1}{2}+r_3\right) \, . \tag{4.4.6}$$

If the two-level system is at its upper level $n_+ = 1$; $n_- = 0$, and the field is in the vacuum state (4.4.5), the vacuum fluctuations (the second term in the r.h.s. of Eq. (4.4.1)) contribute the half of the total intensity of radiation, and the other half is connected with the fluctuations of the dipole moment. When the two-level system is in its ground state ($n_- = 1, n_+ = 0$), the vacuum fluctuations exactly compensate the first term in the r.h.s. of (4.4.1) and the total intensity vanishes. Thus the crucial role of the vacuum fluctuations emerges in the ground state of the matter. The stability of the ground state (i.e. the fact that it does not radiate) is a purely quantum effect which is due to the vacuum fluctuations.

In the general case when both spontaneous and stimulated processes are taken into account, (4.4.1), the change of the ν-th mode photons are obtained from (4.4.1) and (4.4.3)

$$w_\nu = \frac{I_\nu}{\hbar\omega_\nu} = \frac{I_\nu^0}{\hbar\omega_\nu}\left[(n_+ - n_-)\bar{n}_\nu + n_+\right] = \frac{\pi(\vec{e}_1\vec{A}_\nu)^2}{4\hbar\omega_\nu}\left[n_+(\bar{n}_\nu+1) - \bar{n}_\nu n_-\right]\,\delta(\omega_0-\omega_\nu) \, . \tag{4.4.7}$$

It is worthwhile mentioning that we already met expressions similar to Eqs. (4.4.7). The similarity of Eqs. (4.4.7) and (3.9.19) and (3.9.20) shows that irreversible relaxation processes described by Eqs. (3.9.19) and (3.9.20) are due to spontaneous and stimulated emission (absorption) of the phonons.

4.5 Superradiance (small volumes $L^3 << \lambda^3$)

We now consider spontaneous emission of the system of identical two-level molecules located in the volume $L^3 << \lambda^3$, where λ is the wavelength of emitted photons. In the case of the spontaneous emission we use Eq. (4.4.5), $\overline{n}_v = 0$, and substitute it into Eq. (4.3.14). As a result we get the intensity of the spontaneous emission of the system of two-level molecules

$$I = I_0 \left\langle \overline{R_1^2} + \overline{R_2^2} + \overline{R_3} \right\rangle = I_0 \left\langle \overline{\hat{R}^2} - \overline{R_3^2} + \overline{R_3} \right\rangle \ . \tag{4.5.1}$$

Here I_0 is the intensity of spontaneous emission of <u>one</u> two-level molecule.

The square length of the effective spin is

$$\hat{R}^2 = R_1^2 + R_2^2 + R_3^2 \ . \tag{4.5.2}$$

As is known the quantum number corresponding to $\hat{R}^2$ is R: the mean value of $\hat{R}^2$ in the state with definite R is R(R+1). Operator $\hat{R}^2$ commutes with the Hamiltonian (4.3.1) and therefore it is the integral of motion of Eq. (4.3.5). The quantum number corresponding to the operator R_3 is M (M = R, R-1,...,-R).

In the state with definite values of R^2 and R_3 the intensity of the spontaneous emission is, (4.5.1)

$$I = I_0 \left[R(R+1) - M^2 + M \right] = I_0 (R+M)(R-M+1) \ . \tag{4.5.3}$$

It follows from the properties of the effective spin R_3 that quantum numbers M have the upper limit

$$M \le R \ . \tag{4.5.4}$$

Quantum number R in its turn satisfies the relation which follows from (4.3.3),

$$R \le \frac{N}{2} = \frac{N_+ + N_-}{2} \ . \tag{4.5.5}$$

Thus, there are states of the system of two-level molecules located in the volume $L^3 << \lambda^3$, which give spontaneous emission with an intensity proportional to the square of the number of molecules. Such, for example, is the state with $R = \frac{1}{2}N$, M = 0. The intensity of the spontaneous emission from a system in this state is (N >> 1)

$$I = I_0 \frac{1}{2} N \left(\frac{1}{2} N + 1 \right) \approx \frac{1}{4} N^2 I_0 \ . \tag{4.5.6}$$

This is a superradiant state. On the other hand, there are the states in which the system does not radiate the energy at all. An example of this is the state with R = M = 0. We recall that conventional intensity (without taking into account the coherence in the emission) is

$$I = NI_0 \ . \tag{4.5.7}$$

The superradiance was first introduced by Dicke [93] in 1954. Further development of the research in this area has been covered in a number of reviews [94-103].

The phenomenon of the superradiance is not purely quantum effect and can be understood classically. The coherence in the spontaneous radiation can be understood if we recall that each molecule is affected by the field radiated by other molecules. Let us take a system of oscillating charges occupying a region with a linear dimension much less than the length of the emitted wave. In the dipole approximation the radiation damping force acting on each of the oscillators is (see e.g. Landau and Lifshitz [86])

$$\vec{F}_{rad} = \frac{2}{3c^3}\sum_{s=1}^{N} e_s^2 \dddot{\vec{r}}_s \ , \tag{4.5.8}$$

where $\vec{r}_s$ is the displacement of the s-th oscillator; e_s is the charge.

When there is no interaction between the oscillators (apart from the interaction due to the radiation), the coordinates $\vec{r}_k$ are subject to the equations

$$\ddot{\vec{r}}_k + \omega_k^2 \vec{r}_k = \frac{2}{3mc^3}\sum_{s=1}^{N} e_s^2 \dddot{\vec{r}}_s \ . \tag{4.5.9}$$

It follows from this that the interaction due to the radiation of the motion of each oscillator is dependent on the motion of all other oscillators (we do not consider here the change of the frequencies due to the radiation interaction; later on we will consider this problem). In the case when all the oscillators have the same frequency ω_0, and charges $e_k = e$, equation (4.5.9) can be rewritten in the form

$$\ddot{\vec{r}}_k + \omega_0^2 \vec{k}_k = -\gamma_0 \sum_{\ell=1}^{N} \dot{\vec{r}}_\ell \ . \tag{4.5.10}$$

Here we have replaced $\dddot{\vec{r}}$ by $-\omega_0^2 \dot{\vec{r}}_k$, assuming that the width of the emission line, given by $\gamma_0 = 2e^2\omega_0^2/3mc^3$, is much less than the frequency ω_0.

The strength of the radiation emission of the system of oscillators is proportional to the second derivative of the dipole moment of the system

$$\vec{d} = e\sum_{k=1}^{N} \vec{r}_k \ . \tag{4.5.11}$$

Summing Eq. (4.5.10) we get

$$\ddot{\vec{d}} + \omega_0^2 \vec{d} + N\gamma_0 \dot{\vec{d}} = 0 \ . \tag{4.5.12}$$

The solution of Eq. (4.5.12) is approximately

$$\vec{d} = \vec{d}_0 \exp\left[-\frac{N\gamma_0 t}{2}\right] e^{i\omega_0 t} + \text{c.c.} \tag{4.5.13}$$

Therefore the strength of the radiation field has the form

$$\vec{E} = \vec{E}_0 e^{-\frac{N\gamma_0 t}{2}} e^{i\omega_0 t} + \text{c.c.} \tag{4.5.14}$$

This means that a system of N oscillators has a natural linewidth

$$\gamma = N\gamma_0 \ , \tag{4.5.15}$$

where γ_0 is the natural linewidth of a single isolated oscillator.

The intensity of radiation from a system of classical oscillators (in small volume $L^3 << \lambda^3$)

$$I = \frac{2}{3c^3}\left(\ddot{\vec{d}}\right)^2 \ , \tag{4.5.16}$$

is generally not equal to the sum of the intensities from the individual oscillators. In the state when all the oscillators have the same phase ϕ and amplitude d_0,

$$\vec{d}_k = \vec{d}_0 \cos(\omega_0 t + \phi) \ , \tag{4.5.17}$$

the intensity is proportional to N^2. In this example the superradiance (intensity of the radiation is proportional to N^2) is the consequence of the preparation of the system.

4.6 Large sample superradiance

General Maxwell-Bloch equations (4.2.7) and (4.2.8) represent a formulation of the collective spontaneous emission problem valid for any shape or size of the atomic sample. However, in general, these equations are extremely complicate since they involve a summation over all atoms and explicitly or implicitly over all electromagnetic field modes.

An important step in the simplification of the large extent superradiance is to choose the sample having the shape of a long cylinder of length L and radius w, obeying the relations

$$L >> w >> \lambda \ . \tag{4.6.1}$$

This is the so-called pencil-shaped sample case, which has been realized in most experiments (see, e.g., the review by Vrehen and Gibbs [102]). It may be seen that for a long enough sample (along the z direction) one can neglect variation along the transverse x and y directions which amounts to considering the superradiance as being essentially a one-dimensional problem. When one looks more carefully, the situation is somewhat more complicated and the description of superradiance appears to depend upon the Fresnel number

$$F=\frac{\pi w^2}{L\lambda} \quad . \tag{4.6.2}$$

It turned out [97] that the best choice for which one can expect to have a radiation not too different from a "one-dimensional" problem is the one corresponding to

$$F=\frac{\pi w^2}{L\lambda}=1 \quad . \tag{4.6.3}$$

Having in mind the above reservations we will present now a single-mode model for the superradiance [93,21]. We will again start from the Hamiltonian (4.2.4)-(4.2.6), rewriting it in a more convenient form

$$H=\hbar\omega_0 R_3+\sum_{\vec{k}}\hbar\omega_{\vec{k}} a_{\vec{k}}^{+} a_{\vec{k}}+\sum_{\vec{k}}\left(g_{\vec{k}} R_{\vec{k}}^{+} a_{\vec{k}}+g_{\vec{k}}^{+} a_{\vec{k}}^{+} R_{\vec{k}}^{-}\right) \quad , \tag{4.6.4}$$

where $a_{\vec{k}\lambda}^{+}$ $(a_{\vec{k}\lambda})$ is the creation (annihilation) operator for a field quantum (photon) in the mode $(\vec{k},\lambda)$, with wave $\vec{k}$, frequency $\omega_{\vec{k}}$ and polarization $\vec{e}_\lambda$. The coupling constant $g_{\vec{k}\lambda}$ is

$$g_{\vec{k}\lambda}^{*}=\sqrt{\frac{2\pi\hbar}{V\omega_{\vec{k}}}}\left\langle + \,|\, \vec{j}^{+}\cdot\vec{e}_\lambda \,|\, - \right\rangle \quad , \tag{4.6.5}$$

where $\vec{j}^{+}$ is the operator representing current (in the dipole approximation): $\vec{j}^{+}=i(\omega_0\vec{d})$, and $|+>$ and $|->$ are the wave functions of the excited and ground states of the atom, respectively. Here and further on we adopt the notation $\vec{k}\equiv(\vec{k},\lambda)$. The collective $R_{\vec{k}}^{\pm}$ and R_{3k} operators can be expressed in terms of the above (section 3.3), introduced effective spin operators r_{1j}, r_{2j}, r_{3j}, and $r_j^{\pm}=r_{1j}\pm ir_{2j}$.

$$R_{\vec{k}}^{\pm}=\sum_{j=1}^{N} r_j^{\pm}\exp(\pm i\vec{k}\vec{r}_j);\quad R_{3k}=\sum_{j=1}^{N} r_{3j}\exp(i\vec{k}\cdot\vec{r}_j) \quad . \tag{4.6.6}$$

These operators satisfy the commutation relations

$$\left[R_{\vec{k}}^{+}, R_{\vec{k}'}^{-}\right] = 2R_{3(\vec{k}-\vec{k}')}; \qquad \left[R_{\vec{k}}^{\pm}, R_{3}\right] = \mp R_{\vec{k}}^{\pm} \ . \tag{4.6.7}$$

In the single-mode approximation (in which there is only one $R_{\vec{k}}^{\pm}$), a theory can be constructed in complete analogy with the case of a system with dimensions smaller than the wavelength. A role of conserving operator (commuting with the Hamiltonian) R^2, (4.5.2), now plays the operator

$$\hat{R}_{\vec{k}}^{2} = \frac{1}{2}\left(R_{\vec{k}}^{+}R_{\vec{k}}^{-} + R_{\vec{k}}^{-}R_{\vec{k}}^{+}\right) + R_{3}^{2} \ , \tag{4.6.8}$$

which reduces to $\hat{R}^2$ when $\vec{k} = 0$. In the single mode approximation $\hat{R}_{\vec{k}}^{2}$ commutes with *H*, (4.6.4), and is conserved. It was shown by Dicke [93] that the intensity of the radiation of the whole system of N atoms per unit solid angle $d\Omega_k$ in the direction close to axis $\vec{k}$, can be presented as

$$I(\vec{k}) = I_0(\vec{k})\left\langle R_{\vec{k}}^{+}R_{\vec{k}}^{-}\right\rangle \ . \tag{4.6.9}$$

Let us consider two cases – coherent and noncoherent spontaneous emission of the system of the two level molecules. Substituting $R_{\vec{k}}^{\pm}$, (4.6.6), into Eq. (4.6.9), we get

$$I(k) = I_0(\vec{k})\sum_{j,j'}\left\langle r_j^{+}r_{j'}^{-}\right\rangle e^{i\vec{k}(\vec{r}_j-\vec{r}_{j'})} \ . \tag{4.6.10}$$

First, we consider the noncoherent spontaneous emission of the system of two-level molecules. In this case the correlations between various molecules vanish

$$\left\langle r_j^{+}r_{j'}^{-}\right\rangle = \delta_{jj'}\left\langle r_j^{+}r_{j'}^{-}\right\rangle = \delta_{jj'}\left(\frac{1}{2} + r_{j3}\right) \ , \tag{4.6.11}$$

and

$$I = \int d\Omega_{\vec{k}} I(k) = \sum_{j=1}^{N} I_0\left(\frac{1}{2} + r_{j3}\right) \ . \tag{4.6.12}$$

Thus the intensity of the spontaneous emission of N two-level, noncorrelated (noncoherent) molecules is equal to the sum of the intensities of spontaneous emissions of each molecule (4.4.6). In the case when each molecule is in the same state, r_3, then

$$I = NI_0\left(\frac{1}{2} + r_3\right) \ . \tag{4.6.13}$$

Now let us consider the system of correlated coherent two-level molecules. We consider a collection of two-level molecules with non-overlapping wave functions excited by an intense plane-wave pulse characterized by frequency ω_0 and wave vector $\vec{k}$. We assume that this exciting pulse may be described classically, and that it acts on each molecule in the same way except for the time delays due to the finite separations of the molecules from each other. It is well known [93,22] that such a pulse leaves each molecule in a coherent superposition of its upper and lower states. For the ℓ-th molecules, after the passage of the pulse, we can write

$$|\psi_\ell\rangle = e^{i\vec{k}\cdot\vec{r}_\ell/2}\sin\frac{1}{2}\theta\,|+\rangle_\ell + e^{-i\vec{k}\cdot\vec{r}_\ell/2}\cos\frac{1}{2}\theta\,|-\rangle_\ell \equiv |\theta\rangle_\ell \ . \tag{4.6.14}$$

The probability that the molecule has been left in its upper state, equals $\sin^2\frac{1}{2}\theta$. The intensity of the coherently correlated molecules resulted from the pulse is [103]

$$I = I_0\left[\frac{\mu}{4}N^2\sin^2\theta + 1\right] \ . \tag{4.6.15}$$

The quantity μ characterizes the solid angle $\Delta\Omega$ in which the radiation is confined [97]

$$\Delta\Omega = 4\pi\mu \ . \tag{4.6.16}$$

In general μ is a complicated function of the size and shape of the volume in which the two-level molecules are contained. In the case in which the confining volume is a circular cylinder, μ is equal [103]

$$\mu = \frac{6(N-1)}{NG^2H^2}\int_{-1}^{1}\frac{dx(1+x^2)}{(1-x)^2(1-x^2)}\sin^2\left[\frac{1}{2}H(1-x)\right]J_1^2\left[G(1-x^2)^{1/2}\right] \ . \tag{4.6.17}$$

Here J_1 is the Bessel function of the first kind, order 1, and G and H are dimensionless variables

$$G = \frac{2\pi w}{\lambda}; \qquad H = \frac{2\pi L}{\lambda} \ , \tag{4.6.18}$$

where w is a radius of the cylinder and L is its height. μ depends on the Fresnel number (4.6.2). The integral (4.6.17) can be estimated asymptotically for the two limits of a large disk and a long needle. The results for these two limits are [103]

$$\mu = \begin{cases} \dfrac{3}{2G^2}\left(1+\dfrac{\sin^2 H}{H^2}\right), & G >> 1 \text{ and } H << G^2;\ F >> 1 \\ \dfrac{2\pi}{H}, & H >> 1 \text{ and } G << \sqrt{H};\ F << 1 \\ 1-1/N. & H << 1;\ G << 1\ , \end{cases} \tag{4.6.19}$$

where the Fresnel number (4.6.1) can be presented as

$$F = \frac{1}{2}\frac{G^2}{H} \ . \tag{4.6.20}$$

Formulae (4.6.15) – (4.6.19) describe the intensity of the large volume superradiance.

It is expedient to analyze N-dependence of the superradiance intensity [101], $N >> 1$. In the case of small volume the intensity of the superradiance is proportional to N^2 (4.5.4). The intensity of the large volume superradiance (4.6.15) is proportional to μN^2. In the case of large Fresnel numbers $F >> 1$ and $w >> \lambda$, we get from (4.6.19)

$$\mu \approx \frac{3}{8\pi^2}\frac{\lambda^2}{w^2} \qquad (F >> 1)\ . \tag{4.6.21}$$

To get the N dependence we notice that the volume of the sample is

$$V = \pi w^2 L\ . \tag{4.6.22}$$

Therefore the intensity of the superradiance can be presented as

$$I \propto \mu N^2 = \frac{3L\lambda^2 n^2 V^2}{8\pi V} \propto \lambda^2 N^{4/3} n^{2/3}\frac{L^{2/3}}{w^{2/3}} \propto \lambda^2 n^{2/3} N^{4/3}\left(\frac{L}{\lambda}\right)^{1/3} F^{-1/3}\ , \tag{4.6.23}$$

where n is the concentration of the two-level molecules

$$n = \frac{N}{V}\ . \tag{4.6.24}$$

In the case of small Fresnel numbers $F << 1$, one gets

$$I \propto \mu N^2 \propto \lambda^2 n^{2/3} N^{4/3}\left(\frac{L}{\lambda}\right)^{1/3} F^{2/3}; \qquad F << 1\ . \tag{4.6.25}$$

As has been mentioned above, the optimal case is achieved at $F \approx 1$, (4.6.2). This is the pencil-like shape with

$$L = \pi w \left(\frac{w}{\lambda}\right); \qquad L >> w . \tag{4.6.26}$$

In this case, the intensity is

$$I \propto N^{4/3}\left(\frac{L}{\lambda}\right)^{1/3} , \tag{4.6.27}$$

i.e. the intensity I is much larger than what can be achieved when sizes L and w are of the same order of magnitude. In the latter case

$$F = \pi\frac{L}{\lambda} >> 1 , \tag{4.6.28}$$

and I, (4.6.23), is simply proportional to $N^{4/3}$.

Thus the superradiance in large systems (4.6.15), has $N^{4/3}$ – dependence, while the small volume superradiance has N^2 – dependence. A special geometry has to be chosen (the pencil-like case, F ≈ 1) to achieve an essential enhancement over the simple $N^{4/3}$ – dependence.

4.7 Time-development of the superradiance (small volumes)

Now we consider the time evolution of the small volume superradiant systems, described by the Hamiltonian (4.3.1). The time evolution can be found from the conservation law. The total intensity of the spontaneous emission is, (4.5.1),

$$-\hbar\omega_0 \sum_j \frac{d}{dt} < r_{3j} > = -\hbar\omega_0 \frac{d < R_3 >}{dt} = I_0 \left\langle \hat{R}^2 - R_3^2 + R_3 \right\rangle . \tag{4.7.1}$$

Notice that there is a simple solution of this equation in the case of a single molecule, R = $\frac{1}{2}$. In this case

$$- < \dot{r}_3 > = \gamma_0 \left(< r_3 > + \frac{1}{2} \right) . \tag{4.7.2}$$

The solution of this equation is of the form

$$< r_3 > + \frac{1}{2} = n_+ = n_+(0) e^{-\gamma_0 t} , \tag{4.7.3}$$

where $\gamma_0 = I_0/\hbar\omega_0$ is the natural line width of an isolated molecule.

Let us now examine the situation when there is a large number of molecules. We can therefore neglect the mean square deviations R_3 and $\hat{R}^2$ from their mean values and assume that $R^2 >> R_3$. Equation (4.7.1) becomes

$$-\dot{R}_3 = \gamma_0\left(R^2 - R_3^2\right) ; \qquad R^2 = \text{const.} \tag{4.7.4}$$

The last relation follows from the fact that $\hat{R}^2$ commutes the Hamiltonian (4.3.1). To solve equation (4.7.4) we introduce the angle θ between the vector and direction 3 (in the special case of a magnetic moment in a magnetic field, this is the direction of the field)

$$R_3 = R\cos\theta \ . \tag{4.7.5}$$

After substitution of Eq. (4.7.5) into Eq. (4.7.4) one gets

$$\frac{d\theta}{dt} = \gamma_0 R \sin\theta \ . \tag{4.7.6}$$

This equation has been derived by Dicke [93] and solved with the initial condition $\theta(0) = \frac{\pi}{2}$. The solution that satisfies the initial condition

$$\theta(0) = \theta_0 \ , \tag{4.7.7}$$

is of the form [94,21]

$$\tan\frac{\theta}{2} = \tan\frac{\theta_0}{2} e^{\gamma_0 Rt} \ . \tag{4.7.8}$$

From this we can find for the spontaneous emission intensity

$$I = I_0(R^2 - R_3^2) = I_0 R^2 \sin^2\theta = \frac{4R^2 I_0}{(\tan\frac{\theta_0}{2} e^{\gamma_0 Rt} + \cot\frac{\theta_0}{2} e^{-\gamma_0 Rt})^2} \ . \tag{4.7.9}$$

When θ changes from θ_0 to $\pi/2$ (if $\theta_0 < \frac{\pi}{2}$) the amplitude of the radiation rises, reaches a maximum when $\theta = \frac{\pi}{2}$ and then decreases. The spectral intensity is proportional to the square of the modulus of the Fourier component

$$J(\omega) \sim \left| \int_0^\infty \frac{e^{i(\omega_0-\omega)t}}{\tan\frac{\theta_0}{2} e^{\gamma_0 R t} + \cos\frac{\theta_0}{2} e^{-\gamma_0 R t}} dt \right|^2 . \tag{4.7.10}$$

The values of the line width that follow from Eq. (4.7.10) are

$$\Delta\omega = \alpha\gamma_0 R \ , \tag{4.7.11}$$

where in the range of angles θ_0 from 0 to π, α takes up a value from 1 to 2. (With angles θ that are very close to π, when the equation (4.7.6) is inapplicable, Fain [94] has given an additional treatment which shows that $\alpha = 2$.)

If initially at $t = 0$ all spins (two-level molecules) are inverted then

$$R_3 = R = \frac{N}{2} \ , \tag{4.7.12}$$

and $\theta(0) = 0$. In this case Eq. (4.7.6) has the solution at all times

$$\theta(t) = 0 \ . \tag{4.7.13}$$

Of course this solution does not make sense and the approximation (4.7.4), (4.7.5) is not valid at very small angles. One can easily estimate the order of magnitude of the angle θ (or its fluctuations) in the state of full inversion, (4.7.12). Using Eq. (4.7.1) we get

$$I_0 R^2 \sin^2\theta \approx I_0 R^2 \theta_0^2 = I_0\left(R(R+1) - R^2 + R\right) = 2I_0 R \ , \tag{4.7.14}$$

and

$$\theta_0^2 = \frac{2}{R} = \frac{4}{N} \ . \tag{4.7.15}$$

Thus to find the time of the development of superradiance, from the inversion of the population at $t = 0$ to the superradiant state

$$R_3 = 0; \qquad \theta = \frac{\pi}{2} \ , \tag{4.7.16}$$

we have to perform a quantum-mechanical analysis at small angles, (4.7.15). As a result of such an analysis [94] we get for the intensity of the radiation ($N >> 1$)

$$I(t) = I_0 \frac{N^2}{4} \operatorname{sech}^2 (t - t_d)/2\tau_c \ . \tag{4.7.17}$$

The pulse length [see (4.7.11) and (4.7.12)]

$$\tau_c = \frac{\tau_0}{N} \ , \qquad (4.7.18)$$

is smaller by a factor N than the spontaneous emission time of the isolated molecule

$$\tau_0 = \gamma_0^{-1} \ . \qquad (4.7.19)$$

The maximum superradiance intensity

$$I_{max} = I_0 \frac{N^2}{4} \ , \qquad (4.7.20)$$

is reached after the delay time

$$t_d = \tau_c \ln N = \frac{\tau_0}{N} \ln N \ . \qquad (4.7.21)$$

This equation for the delay time was first derived by Fain [94].

Up to now we have discussed electromagnetic interactions only in terms of effects such as attenuation. Electromagnetic interaction leads also to a shift in the energy levels of the system, or to a displacement of the center of the emission line. For a single atom this is the well-known Lamb shift, whose theory was given by Bethe [104]. The frequency shift for the superradiant system was derived by Fain [105]. It has been shown that this shift is much larger than $\tau_c^{-1} = N\gamma_0$.

The important point is that the contribution to this shift is given not only by the resonant modes of the electromagnetic field (like in the calculation of τ_0), but by all modes which give essentially non-zero matrix elements of the spin-field interaction, (4.2.6) and (4.3.4). This means that in the general case, one cannot justify the dipole approximation, (4.3.1), for the whole system. The dipole-dipole van der Waals interaction between the molecules would destroy the development of the coherent spontaneous emission (from $\theta_0 \approx 0$ to $\frac{\pi}{2}$) by not conserving the length of the vector R^2. This was first pointed out by Friedberg, Hartman and Manassah [106].

To avoid misunderstandings, it is worthwhile to stress that the above analysis of this section is correct for the model system described by the Hamiltonian (4.3.1). However, this Hamiltonian cannot be deduced in the consistent way from the Hamiltonian (4.2.6). Nevertheless, as we will see below, the model system, (4.3.1), contains main dependencies and qualitative features of more realistic models.

There are two ways to overcome the destructive dephasing caused by the van der Waals dipole-dipole approximation. One of them is the superradiance in the extended system with volume $V \gg \lambda^3$. In this case only a small volume of $\sim \lambda^3$ gives the contribution to the dephasing, while the coherence is connected with much larger volumes. We will consider this case in the next section. The second way is to put the system in a resonator making use of the cavity enhanced spontaneous emission.

We will consider the case when the system of molecules is located in the volume $V < \lambda^3$ and interacts with only one resonant mode of the resonator. The Hamiltonian (4.3.1) can be approximated as

$$H = \hbar\omega_0 R_3 + \frac{1}{2}\left(p_v^2 + \omega_v^2 q_v^2\right) - \left(\vec{e}_1 \cdot \vec{A}_v\right) q_v R_1 \ ,$$

$$= \hbar\omega_0 R_3 + \hbar\omega_v\left(a_v^+ a_v + \frac{1}{2}\right) - \hbar\alpha R_1\left(a_v^+ + a_v\right) \ . \tag{4.7.22}$$

This Hamiltonian, for spin one half, $R_3 = r_3 = \pm\frac{1}{2}$, and in the rotating wave approximation

$$H = \hbar\omega_0 r_3 + \hbar\omega_v(a_v^+ a_v + \frac{1}{2}) - \frac{\hbar\alpha}{2}(r_- a^+ + r_+ a) \ , \tag{4.7.23}$$

has been suggested by Fain [33,51], and later on by Jaynes and Cummings [191], and became known as the Jaynes-Cummings Hamiltonian. Here a_v^+ and a_v are the creation and annihilation photon operators, and,

$$\alpha = \frac{\left(\vec{e}_1 \cdot \vec{A}_v\right)}{\sqrt{2\hbar\omega_v}} \ , \tag{4.7.24}$$

is the coupling parameter.

The equations for the mean values of the field and spin operators take the form

$$\dot{a}_v = -i\omega_v a_v - \gamma a_v + i\alpha R_1 \ , \tag{4.7.25}$$

$$\dot{a}_v^+ = i\omega_v a_v^+ - \gamma a_v^+ - i\alpha R_1 \ , \tag{4.7.26}$$

$$\dot{R}_1 = -\omega_0 R_2; \quad \dot{R}_2 = \omega_0 R_1 + \alpha\left\langle R_3(a_v^+ + a_v)\right\rangle \ , \tag{4.7.27}$$

$$\dot{R}_3 = -\alpha\left\langle R_2(a_v^+ + a_v)\right\rangle \ . \tag{4.7.28}$$

Here we have introduced the decay constant of the cavity mode

$$\gamma = \frac{1}{2}\frac{\omega_v}{Q} , \tag{4.7.29}$$

and Q is the quality factor.

In the "semiclassical" ("mean field") approximation (in which one can neglect the dispersions of R_1, R_2, R_3), Eqs. (4.7.25 – 4.7.28) can be reduced to ($\omega_0 = \omega_v = \omega$)

$$\ddot{R}_3 + \frac{\alpha^2 a_0^2}{2} \cdot R_3 - \frac{\dot{a}_0}{a_0} R_3 = 0 \ , \tag{4.7.30}$$

where a_0 is the slowly varying amplitude of a_v

$$a_v = a_0 e^{-i\omega t} \ . \tag{4.7.31}$$

The general solution of this equation, as it can be checked by direct substitution, is of the form

$$R_3 = R\cos\theta \ ; \qquad \theta = \frac{\alpha}{\sqrt{2}} \int_0^t a_0(t_1)dt_1 + \theta_0 \ . \tag{4.7.32}$$

The equation giving the energy balance of the molecular + field system has the form

$$\dot{R}_3 + \dot{n} + 2\gamma(n - n_0) = 0 \ . \tag{4.7.33}$$

For the sake of simplicity let us look at the case when $n >> n_0$ (the equilibrium value of the photon number n). Then from (4.7.32) and (4.7.33) we obtain the equation for the angular variable θ

$$\ddot{\theta} + \frac{\omega}{2Q}\dot{\theta} - \Omega_N^2 \sin\theta = 0 \ , \tag{4.7.34}$$

where Ω_N is the collective Rabi frequency

$$\Omega_N^2 = \frac{\alpha^2 R}{2} = \frac{\alpha^2 N}{4} \ , \tag{4.7.35}$$

(the latter equation is valid for $R = \frac{N}{2}$). Equation (4.7.34) was first derived by Fain [51] in 1959 (see also [21]). The problem of the collective emission in the cavity of the resonator is isomorphous to the problem of the falling of a nonlinear pendulum with viscous damping.

For $\Omega_N >> \omega/Q$ the equation (4.7.34) exhibits decaying oscillatory behavior. In the overdamped regime

$$\Omega_N < \omega/Q \ , \tag{4.7.36}$$

equation (4.7.34) reduces to a first-order derivative equation

$$\frac{d\theta}{dt}=\frac{2\Omega_N^2}{\omega}Q\sin\theta=\frac{1}{2T_R}\sin\theta=\frac{\alpha^2 Q}{2\omega}R\sin\theta\ . \tag{4.7.37}$$

This equation is isomorphous to Eq. (4.7.6) and thus may be used to describe the process of collective spontaneous emission in the resonator. The only difference is that instead of the constant τ_0^{-1}, (4.7.19), we have

$$\tilde{\tau}_0^{-1}=\frac{\alpha^2 Q}{2\omega}\ . \tag{4.7.38}$$

It is easy to check that this enhanced spontaneous emission rate [107] is approximately Q times larger than the spontaneous emission rate free space. This circumstance justifies the neglect of the dipole-dipole dephasing as soon as

$$Q>\frac{\lambda}{L}\ ,$$

where $\sim L^3$ is the volume of the sample. The cooperative decay is thus enhanced, whereas the van der Waals dephasing is not modified.

4.8 Time-development of the superradiance (large volumes)

In the single-mode approximation (in which there is only one definite mode $R_k^{\pm}$ is taken into account) a theory of time evolution of the superradiance can be constructed in complete analogy with the case of a system with dimensions smaller than the radiated wavelength. A role of conserving operators similar to $\hat{R}^2$ plays operator $\hat{R}_k^2$, (4.6.8).

Now we can use an analog of the conservation law, (4.7.1),

$$-\hbar\omega_0\dot{R}_3=I(\vec{k})\ . \tag{4.8.1}$$

Using Eq. (4.6.14) and assuming $N>>1$ we define θ by the relation (assuming that θ is not too small)

$$R_3=|R_k|\cos\theta\ . \tag{4.8.2}$$

Then we get from Eqs. (4.6.9) and (4.8.1), and the conservation of $\hat{R}_k^2$, (4.6.8)

$$\frac{d\theta}{dt}=\frac{1}{2T_R}\sin\theta\ . \tag{4.8.3}$$

where we have defined the characteristic superradiance time

$$T_R = \frac{\tau_0}{N\mu} = (N\gamma_0\mu)^{-1} \quad . \tag{4.8.4}$$

Eq. (4.8.3) is isomorphous with Eq. (4.7.6) ($R = \frac{N}{2}$). Here, $\tau_0 = \gamma_0^{-1}$ is the spontaneous emission time of the isolated molecule in free space. Parameter μ is defined by Eqs. (4.6.16) and (4.6.17) (a circular cylinder).

The delay time of the superradiance can be found similarly to the case of small volumes (performing quantum analysis for θ close to 0; see Ref. 94)

$$t_d = T_R \ln N = \tau_0 \ln N / N\mu \quad . \tag{4.8.5}$$

When N is large and the system has dimensions which are small compared with λ, then $\mu = 1$, (4.6.19), and Eq. (4.8.3) agrees with the result found by Dicke [93], while Eq. (4.8.5) reduces to the result found by Fain [94].

The model which has been used for the large volume superradiance time development is sometimes called the "mean-field" theory. It is important to emphasize that the mean-field model of large sample superradiance overlooks the propagation and the field non-uniformly, which it induces in the sample. This model is not justified by any physical approximation and its only merit is the simplicity and symmetry of the equations. However, in spite of the fact that this model neglects important propagation effects, qualitatively it retains some of the essential features of large sample superradiance, as well as some important quantitative dependences.

The propagation effects may play quite an essential role, which is clear from the analysis performed by Arecchi and Courtens [108]. As follows from (4.8.4) T_R can be very small provided $N\mu$ is large enough. On the other hand, the enhanced decay rate of (4.8.4) will be realistic only if the entire sample is shorter than cT_R so that all atoms in the sample are exposed to the radiation from all others before the decay process is completed. Therefore we have to know the appropriate decay rate Γ_c for a sample of size larger than cT_R. The number of atoms (density ρ) which can be covered by the radiation of other atoms during the decay time Γ_c^{-1} is given by

$$N_c = c\Gamma_c^{-1}A\rho \quad . \tag{4.8.6}$$

Here A is the cross-section of the "pencil". On the other hand these atoms superradiate with the rate given by the first-order perturbation theory (4.8.4)

$$\Gamma_c = N_c\gamma_0\mu \quad . \tag{4.8.7}$$

These two relations (4.8.6) and (4.8.7) can be solved self-consistently for Γ_c and N_c

$$N_c = \sqrt{\frac{cA\rho}{\gamma_0\mu}} \quad , \tag{4.8.8}$$

and

$$\Gamma_c = \sqrt{cA\rho\gamma_0\mu} \quad . \tag{4.8.9}$$

Here Γ_c^{-1} is defined as the cooperative time [108]. N_c is the maximum cooperation number [108], i.e. the maximum number of atoms that can cooperate to superradiant emission in a particular experimental situation. Assuming for instance, that a sample could be prepared in the given initial condition and with $N > N_c$, all atoms emit, of course, but only N_c cooperate. In this case the emission rate is not given by (4.8.4) but by (4.8.9). The maximum coherence length $c\Gamma_c$ is independent of the sample geometry and depends only on the transition parameters and on the density.

The estimates of Ref. [108] show that $c\Gamma_c$ are typically 10^4 cm in NMR (at 1 MHz), 10 cm in ESR (at 10 GHz), and 0.1 cm for the 6993 Å transition is ruby with 0.05% Cr^{3+} concentration. The limited cooperation becomes, therefore, of prime importance in optical experiments.

To take into account the propagation [109-114] effects we have to use the Bloch-Maxwell equations (4.2.7)-(4.2.8). We will write them in a different form [97] (see also Ref. 87)

$$\frac{\partial N}{\partial t} = \frac{i}{\hbar}\left[\vec{E}^+(\vec{x},t) + \vec{E}^-(\vec{x},t)\right]\left[\vec{P}^+(\vec{x},t) - \vec{P}(\vec{x},t)\right] \ ,$$

$$\frac{\partial \vec{P}^+(\vec{x},t)}{\partial t} - i\omega_0 \vec{P}^+(\vec{x},t) + 2i\frac{d^2}{\hbar}\vec{\varepsilon}_a\left\{\vec{\varepsilon}_a\left[\vec{E}^+(\vec{x},t) + \vec{E}^-(\vec{x},t)\right]\right\} N \ . \tag{4.8.10}$$

Here the electric field and polarization are divided into positive and negative frequency parts, $N = 2r_3(\vec{x})$ is the inversion population difference and $\vec{\varepsilon}_a$ is the unit vector of the atom's dipole moment polarization. The Maxwell equation describing the propagation of electromagnetic waves in the medium with the polarization $\vec{P}(\vec{x},t)$ has to be added.

In order to describe the pencil-shaped sample collective spontaneous emission one assumes that the electric field and atomic polarization field can be expressed as products of a fast varying exponential (with time and space frequencies ω_o and k_o) and slow varying enveloped E(z,t) and P(z,t)

$$E^{\pm} = E^{\pm}(z,t)\exp\left[\mp i(\omega_o t - k_o z)\right] \ ,$$

$$P^{\pm} = P^{\pm}(z,t)\exp\left[\pm i(\omega_o t - k_o z)\right] \ . \tag{4.8.11}$$

Neglecting $\frac{\partial E}{\partial t}$ and $\frac{\partial P}{\partial t}$, when compared to $\omega_o E$ or $\omega_o P$, and $\frac{\partial E}{\partial z}$ when compared to $k_o E$ one gets (after dropping fast oscillating terms, the rotating wave approximation) the following Maxwell-Bloch equations for the atom and field envelope operators

$$\frac{\partial N}{\partial t} = \frac{1}{\hbar}[P^+E^+ - E^-P^-] \frac{\partial P^+}{\partial t} = \frac{2id^2}{\hbar} E^-N \ ,$$

$$\left(\frac{1}{c}\frac{\partial}{\partial t} + \frac{\partial}{\partial z}\right)E^+ = \frac{i\omega_o}{2\varepsilon_o c}P^- \ . \tag{4.8.12}$$

These equations take a simpler form, if one replaces the variables z and t by z and the retarded time $\tau = t - z/c$

$$\frac{\partial N}{\partial \tau} = \frac{i}{\hbar}[P^+E^+ - E^-P^-]; \qquad \frac{\partial P^+}{\partial \tau} = \frac{2id^2}{\hbar}E^-N \ , \tag{4.8.13}$$

$$\frac{\partial}{\partial z}E^+ = \frac{i\omega_o}{2\varepsilon_o c}P^- \ . \tag{4.8.14}$$

The classical equations describing the one-dimensional propagation model of superradiance can be easily deduced from these quantum equations. We have just to replace the operators by the corresponding classical quantities. The resulting equations are conveniently solved [111, 112] by introducing a time and space varying tipping angle $\theta(z,t)$

$$\frac{2d}{\hbar}E_{c\ell} = e^{i\phi}\frac{\partial\theta}{\partial t} \ , \tag{4.8.15}$$

$$\frac{\partial^2\theta}{\partial z \partial \tau} = \frac{1}{LT_R}\sin\theta \ , \tag{4.8.16}$$

with T_R being defined by Eq. (4.8.4). We can also write Eq. (4.8.16) in a dimensionless form as:

$$\frac{\partial^2\theta}{\partial(z/L)\partial(\tau/T_R)} = \sin\theta \ . \tag{4.8.17}$$

This equation, known as the Sine-Gordon one, has been introduced in the context of the superradiance in Refs. [111, 112]. Equation (4.8.17) describes the flipping of the local tipping angle $\theta(z,t)$ from a value close to zero at $t = 0$ to the value $\theta = \pi$ at $t = \infty$. The initial conditions to choose for solving these equations are

$$\theta(z, \tau = 0) = \theta^i \ , \tag{4.8.18}$$

$$\frac{\partial\theta(z, \tau = 0)}{\partial \tau} = 0 \ . \tag{4.8.19}$$

An inspection of Eq. (4.8.17) shows that θ is in fact a function of the single dimensionless quantity

$$q = 2\left(\frac{z}{L}\frac{\tau}{T_R}\right)^{\frac{1}{2}} . \tag{4.8.20}$$

The Sine-Gordon equation can be transformed into

$$\theta''(q) + \frac{1}{q}\theta'(q) - \sin\theta = 0 , \tag{4.8.21}$$

with the initial conditions

$$\theta(0) = \theta^i ; \qquad \theta'(0) = 0 . \tag{4.8.22}$$

Once $\theta(q)$ is numerically obtained, one finds $E_{c\ell}(z,t)$ with the help of Eq. (4.8.15) and the emitted intensity at the end of the medium is given by

$$I_{cl}(\tau) = E^*_{cl}(L,\tau)E_{cl}(L,\tau) .$$

The main differences of these solutions with the mean-field model are:

(a) The occurrence of several emission maxima (the so-called superradiance ringings [111]).

(b) Not only delay but also the shape of single shot trajectories depend upon the tipping angle. In particular the ringing contrast depends on θ^i with more important ringings obtained for small θ^i values [111].

(c) Delay time. Following [111] we notice that the maximum of the first emission ringing corresponds to the time when $\theta(z,L) = 1$. To determine the delay time one solves the Sine-Gordon equation for $0 < \theta < 1$, by assuming that $\sin\theta \approx \theta$

$$\theta'' + \frac{1}{q}\theta' - \theta = 0 . \tag{4.8.23}$$

The solution of this equation satisfying initial conditions (4.8.18) and (4.8.19) is

$$\theta(q) = \theta^i J_o(iq) . \tag{4.8.24}$$

The corresponding emission delay τ_d is thus given by

$$J_o\left[2i(\tau_D/T_R)^{\frac{1}{2}}\right] \approx 1/\theta^i . \tag{4.8.25}$$

Replacing the function J_o by its approximate value

$$J_o(iq) = \frac{1}{\sqrt{2\pi q}} e^q \quad . \tag{4.8.26}$$

Eq. (4.8.25) can be rewritten as

$$\tau_d = \frac{1}{4} T_R \left[\mathrm{Log}(A/\theta^i) \right]^2 \quad , \tag{4.8.27}$$

where $A = 2\left(\pi^2 \tau_d / T_R\right)^{\frac{1}{4}}$ can be taken as a constant (of order 10). The average delay τ_D obtained using Eq. (4.7.15) is

$$\theta_i = \frac{2}{\sqrt{N}} \quad , \tag{4.8.28}$$

$$\tau_d = \frac{1}{4} T_R \log\left(\frac{A}{2} \sqrt{N} \right) \quad . \tag{4.8.29}$$

Comparing the "one-dimensional" propagation and the mean-field mode, one notices that delays given respectively by formulae (4.7.21), (4.8.5) and (4.8.29) are not too different from each other, the ratio between them being a logarithmic factor slowly varying with N. For large atom numbers ($N > 10^6$) the delay in the one dimensional propagation model is somewhat larger than the one in the mean-field mode. For example, the ratio between the two delays is equal to 1.5 for $N = 10^9$ ($\tau_d \approx 21\ T_R$ in the mean field model and 31 T_R in the one-dimensional propagation model).

The concept of collective spontaneous emission was formulated 45 years ago [93]. Since then the superradiance has been developed into a quite distinct field of research. Quantum theory of irreversible processes must include the phenomenon of the superradiance, as far as the relaxation due to the interaction with photons is taken into account. In this chapter we presented only basics of the superradiance theory.

The field itself expanded in various directions – semiconductors [115-118], free electron systems [119, 120], nuclear spin systems [121, 122], two-components systems [123, 124], excitons [125, 126] and laser-driven time delayed collective emission [101, 127]. It is not our intention to analyze the intensive development of the theory of collective spontaneous emission.

CHAPTER V

MEMORY EFFECTS IN RELAXATION PROCESSES. EXACT SOLUTIONS.

Most of the results presented in this chapter are exact. The approximate ones are treated in a consistent way, and compared with exact solutions. Two simple models are considered: the spin-boson system – a spin interacting with a phonon bath, and the oscillator-boson system. The exact solution, for the latter system, provides opportunities to assess the rotating wave, the Markovian and the weak coupling approximations. The spin-boson system has been extensively treated by Legget et al [137]. In this chapter we mainly concentrate on the existence of dissipationless regimes in a system having a continuum of states, and generally exhibiting the irreversible behavior. An example of such a system is the spin-boson system. It has been shown that in a certain range of parameters, the system may reveal dissipationless phenomena, such as nondecaying quantum beats, and the non-zero asymptotic ($t \to \infty$) probability to remain in the excited state. (Usually, in the relaxation due to the spontaneous emission of bosons, the probability to remain in the excited state vanishes when $t \to \infty$.) Most of the results in this chapter are obtained assuming zero absolute temperature ($T = 0$). This means that condition (2.3.14) does not lead to the factorization approximation. However, most of the solutions in this paper are exact and do not require the factorization approximation. They take into account the memory effects which were neglected in the Markovian approximation.

5.1 Time development of quantum systems: general relations

In Chapter II various approximate approaches to irreversible processes have been presented. In this chapter we consider a number of exact relations and solutions.

Consider a system described by the Hamiltonian

$$H = H_o + V \ , \tag{5.1.1}$$

where the energy spectrum of the unperturbed Hamiltonian H_o contains both discrete and continuous energy levels and V is the perturbation energy. Let S be a unitary matrix which connects eigenstates ψ'_L of the Hamiltonian H with eigenstates ψ_v of the Hamiltonian H_o

$$\psi'_L = \sum_v S_{vL}\psi_v \ , \tag{5.1.2}$$

or equivalently

$$S_{vL} = \langle \psi_v \mid \psi'_L \rangle \ . \tag{5.1.3}$$

The unitary matrix S performs the diagonalization of the Hamiltonian H. In the representation of ψ_v, eigenfunctions of the Hamiltonian H_o, the Hamiltonian H_o is diagonal, while the Hamiltonian H is off-diagonal. In the representation of function ψ'_L

$$H\psi'_L = E'_L\psi'_L \quad , \tag{5.1.4}$$

the Hamiltonian H, including the perturbation energy, is diagonal. Thus the arbitrary operator F and density matrix, in a new representation, have the form (see (1.2.10), and (1.2.14))

$$F' = S^{-1}FS \quad , \tag{5.1.5}$$

$$\rho' = S^{-1}\rho S \ . \tag{5.1.6}$$

In the new representation, the von Neumann equation for the density matrix has the form

$$\dot{\rho}'_{LM} = -i\omega'_{LM}\rho'_{LM} \quad , \tag{5.1.7}$$

i.e.,

$$\rho'_{LM} = \rho'_{LM}(0)e^{-i\omega'_{LM}t} \quad , \tag{5.1.8}$$

where

$$\omega'_{LM} = \left(E'_L - E'_M\right)/\hbar \quad , \tag{5.1.9}$$

where E'_L and E'_M are eigenvalues of the Hamiltonian H, (5.1.1).

We will be interested in the time development of the density matrix ρ in the representation of the unperturbed states ψ_v. For this purpose we can use the relation inverse to (5.1.6)

$$\rho = S\rho' S^{-1} \ . \tag{5.1.10}$$

We find, from (5.1.10) and (5.1.8) that

$$\rho_{uv}(t) = \sum_{LM} S_{uL}\rho'_{LM}(0)S^*_{vM}e^{-i\omega'_{LM}t} \ . \tag{5.1.11}$$

To express $\rho'_{LM}(0)$ through the initial conditions for the unperturbed density matrix ρ(0), we use relation (5.1.6)

$$\rho'_{LM}(0)=\sum_{u'v'} S^{*}_{u'L}\rho_{u'v'}(0)S_{v'M} \quad . \tag{5.1.12}$$

Substituting this relation into Eq. (5.1.13) we get

$$\rho_{uv}(t)=\sum_{L,M,u',v'} S_{uL}S^{*}_{u'L}\rho_{u'v'}(0)S_{v'M}S^{*}_{vM}e^{-i\omega'_{LM}t} \quad . \tag{5.1.13}$$

It is easy to check that the r.h.s. of Eq. (5.1.13) satisfies the initial condition $\rho_{uv}(0)$. Putting $t=0$ we get

$$\rho_{uv}(0)=\sum_{Lu'}\left(S_{uL}S^{*}_{u'L}\right)\sum_{M'v'}\left(S_{v'M}S^{*}_{vM}\right)\,\rho_{u'v'}(0) \quad . \tag{5.1.14}$$

Due to the unitarity of matrix S, (1.2.9), we get

$$\sum_{L} S_{uL}S^{*}_{u'L}=\delta_{uu'}\;; \qquad \sum_{M} S_{v'M}S^{*}_{vM}=\delta_{vv'} \quad . \tag{5.1.15}$$

Substituting these relations into (5.1.14) we obtain that the r.h.s. of Eq. (5.1.14) becomes $\rho_{uv}(0)$. Thus the r.h.s. of Eq. (5.1.13) is the superposition of exponents $\exp(-i\omega'_{LM})t$, and therefore satisfies the von Neumann equation (5.1.7) and it also satisfies the initial condition $\rho_{uv}(0)$. Therefore, Eq. (5.1.13) represents the exact time dependence of the density matrix $\rho_{uv}(t)$.

Indices u, v characterise discrete (including the ground state) and continuous energies of the unperturbed Hamiltonian H_o. On the other hand, among eigenvalues E'_L there may also be discrete E'_n (including the ground state E'_g) and continuous energies E'_α. Then density matrix ρ_{uv} (in the presentation of the Hamiltonian H_o) may be presented as

$$\begin{aligned}\rho_{uv}(t)=&\sum_{n,m,u'v'} S_{un}S^{*}_{u'n}\rho_{u'v}(0)S_{v'm}S^{*}_{vm}e^{-i\omega'_{nm}t}\\ &+\sum_{\alpha nu'v'}\{S_{un}S^{*}_{u'n}\rho_{u'v'}(0)S_{v'\alpha}S^{*}_{v\alpha}e^{i\omega'_{\alpha n}t}\\ &+S_{u\alpha}S^{*}_{u'\alpha}\rho_{u'v'}(0)S_{v'n}S^{*}_{vn}e^{-i\omega'_{\alpha n}t}\}\\ &+\sum_{\alpha\beta u'v'} S_{u\alpha}S^{*}_{u'\alpha}\rho_{u'v'}(0)S_{v'\beta}S^{*}_{v\beta}e^{-i\omega'_{\alpha\beta}t} \quad .\end{aligned} \tag{5.1.16}$$

A summation over indices α, β could be transformed into an integration. This means that the last three terms in (5.1.16) become the Fourier integrals which will decay to zero or to some constant term (see section 1.8), as $t\rightarrow\infty$. On the other hand, the first term in the r.h.s. of Eq. (5.1.16) does not have a limit (for $n\neq m$). It is not irreversible. Thus, existence of discrete states of the total Hamiltonian (5.1.1) (apart from the ground state) is a condition of the asymptotic dissipationless behavior of the system described by Hamiltonian (5.1.1).

5.2 General criteria for emerging of dissipationless regimes

Macroscopic systems usually have dissipative properties and irreversible behavior. An energy concentrated initially at a certain point, or small volume, is dissipated over the whole system. The electric current circuit is characterized by the resistance. An electromagnetic excitation in the resonator cavity decays to the equilibrium value. On the other hand, it is well known that a generally dissipative system, in a certain range of parameters of the system, may reveal dissipationless phenomena, such as superconductivity, superfluidity, local vibrations, etc.

Now we will consider general conditions of emerging of the dissipationless states in the dissipative system having a continuum of states [128]. In the general case the system described by the diagonalized Hamiltonian (5.1.1), may have either a continuous energy spectrum only, or both discrete and continuum energy states, while the energy spectrum of the unperturbed Hamiltonian contains discrete and continuum energy levels ε_a

$$H_o|a> = \varepsilon_a|a> \ . \tag{5.2.1}$$

As has been noticed, the existence of discrete states of the total Hamiltonian is a condition for the asymptotic dissipationless behavior of the system described by the Hamiltonian (5.1.1). Therefore, let us concentrate on the conditions of the existence of such states.

Following the usual procedure [129], we define the Green function

$$G_a(E) = \langle a \,|\, (E-H)^{-1} \,|\, a > \rangle = \sum_L \frac{|< a \,|\, L >|^2}{E - E_L'} = \frac{1}{E - \varepsilon_a - R_a(E)} \ , \tag{5.2.2}$$

$$R_a(E) = < a \,|\, R \,|\, a > \ .$$

Here the level shift operator R is determined by the equation

$$R = V + V \frac{1-\Lambda_a}{E - H_o} R \ , \tag{5.2.3}$$

where $\Lambda_a = |a >< a|$ is the projection operator, and it is assumed that R is to operate on the state |a>. Energy levels E_L' of the Hamiltonian H and function $R_a(E_L')$ can be found from the equations

$$E_L' = \varepsilon_a + R_a\left(E_L'\right) \ , \tag{5.2.4}$$

$$R_a(E_L') = \sum_{a'(\neq a)} \frac{<a|V|a'><a'|V|a>}{E_L' - \varepsilon_{a'}}$$

$$+\sum_{a',a'',a'''} \frac{<a|V|a'><a'|V|a''><a''|V|a'''><a'''|V|a>}{(E_L' - \varepsilon_{a'})(E_L' - \varepsilon_{a''})(E_L' - \varepsilon_{a'''})} + \ldots \quad (5.2.5)$$

Coefficients $|S_{aE_L'}|^2 = |<a|E_L'>|^2$ of the unitary matrix S can be found from Eq. (5.2.2). According to Eq. (5.2.2) we can write

$$G_a(E_d + \delta) = \frac{|<a|d>|^2}{\delta} + \sum_\alpha \frac{|<a|\alpha>|^2}{E_d - E_\alpha + \delta} + \sum_{d'(\neq d)} \frac{|<a|d'>|^2}{E_d - E_{d'} + \delta} \ .$$

Here $|d>$ are discrete states of the exact Hamiltonian (5.1.1), these states are assumed to be nondegenerate $E_d \neq E_{d'}$, and $|\alpha>$ are exact continuous states. The second and third terms are finite when $\delta \to 0$. The second term is transformed into a singular integral, and using (5.2.2) we get

$$|<a|d>|^2 = \lim_{\delta \to 0} \delta G(E_d + \delta) = \frac{\delta}{\delta(1 - R_a'(E_d))} \ ,$$

and

$$|S_{ad}|^2 = <a|d>|^2 = \frac{1}{1 - R_a'(E_d)} \ . \quad (5.2.6)$$

For simplicity it has been assumed that expansion (5.2.5) contains only even powers of V. In the examples which will be given below this assumption is valid. Generally, Eq. (5.2.5) contains both odd and even terms. The generalization to this case can be performed as well.

Now let us make certain assumptions about the energy spectrum of the unperturbed system ε_a, and matrix elements $V_{aa'}$. The onset of the continuous part of the unperturbed energy spectrum is assumed to start at $\varepsilon_a = 0$, while the discrete excited energies $\varepsilon_n \geq 0$. (It can be shown, see (5.2.8), that for $\varepsilon_n < 0$ the combined system always has discrete energy levels.) We also assume that the numerators on the r.h.s. of Eq. (5.2.5) are positive. These assumptions are satisfied by the examples presented below.

While the unperturbed energy spectrum contains both discrete and continuum states, the spectrum of the exact energy levels E_L' may have either continuum states only (apart from the ground state), or both continuum and discrete states. In the former case, the system is dissipative, while in the latter case the system reveals asymptotically nondissipative behavior (the first term in the r.h.s. of Eq. (5.1.16)).

It can be shown [130-132] that the exact continuum states have positive energy levels $E_L' \geq 0$. On the other hand the existence of negative excited discrete energy levels is a necessary condition for the nondissipative behavior of the system.

The exact negative state E_d=-E_o, $E_o > 0$ corresponding to the excited unperturbed state m, can be found from Eqs. (5.2.4) and (5.2.5)

$$\varepsilon_m + E_o = -R_m(-E_o) = \sum_{a'(\neq m)} \frac{|V_{ma'}|^2}{E_o + \varepsilon_{\alpha'}} + \sum_{\substack{a',a'',a''' \\ (\neq m)}} \frac{V_{ma'}V_{a'a''}V_{a''a'''}V_{a'''m}}{(E_o + \varepsilon_{a'})(E_o + \varepsilon_{a''})(E_o + \varepsilon_{a'''})} + \dots \quad . \tag{5.2.7}$$

In a sense, the emergence of the discrete (negative) energy level located below the continuum, may be presented as the Lamb shift "pushing" the energy level below the continuum.

Utilizing the above-mentioned properties of matrix elements of V, one can easily verify that $-R_n(-E_o)$ is a monotonically declining function of E_o. The left hand side of Eq. (5.2.7) is a linearly increasing function of E_o. This means that the positive root of E_o (negative E_d) exists, provided the following condition is satisfied

$$\varepsilon_m < \varepsilon_m^c = -R_m(0) = \sum_{a'(\neq m)} \frac{|V_{ma'}|^2}{\varepsilon_{a'}} + \sum_{\substack{a',a'',a''' \\ (\neq m)}} \frac{V_{ma'}V_{a'a''}V_{a''a'''}V_{a'''m}}{\varepsilon_{a'}\varepsilon_{a''}\varepsilon_{a'''}} + \dots \quad . \tag{5.2.8}$$

The existence of nonvanishing $S_{md} \neq 0$, (5.2.6), provides a condition of nondissipative quantum oscillations.

It should be emphasized that in the general case the above relations do not coincide with the conventional perturbation theory, where in the zeroth approximation one derives from Eqs. (5.2.4) and (5.2.5), $E_L' = \varepsilon_a$, and in the second approximation one puts $E_L' = \varepsilon_a$ on the r.h.s. of Eq. (5.2.4). This term has to be close to $E_L' = \varepsilon_a$. On the other hand, by putting $E_o = 0$ into the r.h.s. of Eq. (5.2.4) we do not require the smallness of the term $\sum |V_{ma'}|^2/\varepsilon_{a'}$ in comparison with $E_o = 0$, and ε_m may be essentially different from zero. The present analysis may be developed by comparing terms on the r.h.s. of Eq. (5.2.8), i.e.

$$\sum_{\substack{a',a'',a''' \\ (\neq m)}} \frac{V_{ma'}V_{a'a''}V_{a''a'''}V_{a'''m}}{\varepsilon_{a'}\varepsilon_{a''}\varepsilon_{a'''}} \ll \sum_{a'(\neq m)} \frac{|V_{ma'}|^2}{\varepsilon_{a'}} \quad . \tag{5.2.9}$$

One has to distinguish two cases: (a) $\lim_{E_o \to 0} -R_m(-E_o)$ exists, i.e., is finite; then Eq. (5.2.7) has a solution if, and only if, condition (5.2.8) is satisfied; (b) $R_m(-E_o)$ diverges at $E_o \to 0$ – then there is a solution for all ε_n. In case (a), employing the fact that $R_m(-E_o)$ is a monotonically declining function of $E_o > 0$, it is easy to find that the root $E_o = -E_d$ increases monotonically from 0 to E_o^{max} satisfying the equation

$$E_o^{max} = -R_m\left(-E_o^{max}\right) \; , \tag{5.2.10}$$

when ε_m decreases from ε_m^c to zero. At the same time $|S_{md}|^2$, (5.2.6), increases monotonically to its maximum value at $\varepsilon_m = 0$, when ε_m decreases from ε_m^c to zero.

Thus, it has been shown that in quite a general case the system described by the Hamiltonian (5.1.1) may contain a range of its excited unperturbed states $\varepsilon_m < \varepsilon_m^c$, in which the system reveals the non-dissipative behavior. It should be mentioned, however, that the nondissipative behavior may be observed, provided $|S_{md}|^2$ has an essentially nonzero value. In the case when $|S_{md}|^2 << 1$, the nondissipative features are practically unobservable.

5.3 The configuration interaction between one discrete state and the continuum of the states

Generally one can find matrix elements of the unitary matrix (5.1.3) and (5.2.6) and exact energy levels E_L', (5.1.4), only in a certain approximation. But there exists a specific example when a unitary transformation can be performed exactly. This example was considered in detail by Rice [130], and Fano [132] (see also [131]) in the context of the configurational interaction in atoms and molecules.

Electronic states of atoms and molecules are usually classified as belonging to various configurations. This classification is an approximate one, based on the independent particle approximation. The exact stationary states can be generated by the "configurational interaction", i.e., by the terms of the Hamiltonian that are omitted in the independent particle approximation. The mixing of a configuration belonging to a discrete spectrum with a continuous one may give rise to the phenomenon of the autoionization.

We consider the case when the whole system has one discrete $|m\rangle$ state degenerate with the continuous states $|\gamma\rangle$. The Hamiltonian of the system has the form

$$H = H_o + V \ . \tag{5.3.1}$$

We assume that the Hamiltonian H has been diagonalized in the continuum states. This means that matrix elements

$$V_{\gamma\gamma'} = 0 \ , \tag{5.3.2}$$

while the only nonvanishing matrix elements of the perturbation energy are $V_{m\gamma}$. In this case the exact energy levels E of the whole system are determined by equations (5.2.4) and (5.2.5)

$$E = \varepsilon_m + R_m(E) \ , \tag{5.3.3}$$

$$R_m(E) = \sum_\gamma \frac{|V_{m\gamma}|^2}{E - E_\gamma} \ , \tag{5.3.4}$$

while the unitary matrix elements have the form (5.2.6)

$$|S_{m\gamma}|^2 = \left[1 + \sum_\gamma \frac{|V_{m\gamma}|^2}{(E - E_\gamma)^2}\right]^{-1} . \tag{5.3.5}$$

As previously we have chosen the zero of the unperturbed energy levels at the onset of the continuum E_γ, while the unperturbed discrete energy E_m is non-negative

$$E_m \geq 0 . \tag{5.3.6}$$

First we will find the discrete state satisfying Eqs. (5.3.3) and (5.3.4). The existence of such a discrete state, in the context of the configuration interaction, has been discussed in several papers [133-136]. The exact discrete state of the system described by the Hamiltonian (5.3.1), has the negative energy

$$E_d = -E_o ; \quad E_o > 0 . \tag{5.3.7}$$

According to Eqs. (5.2.7), (5.3.3) and (5.3.4), E_o satisfies the equation

$$E_n + E_o = \sum_\gamma \frac{|V_{m\gamma}|^2}{E_o + E_\gamma} . \tag{5.3.8}$$

This equation has a positive root (and negative $E_d = -E_o$), provided condition (5.2.8) is satisfied

$$E_m < E_m^c = \sum_\gamma \frac{|V_{m\gamma}|^2}{E_\gamma} . \tag{5.3.9}$$

The "weight" of this state equals

$$|S_{md}|^2 = \left[1 + \sum_\gamma \frac{|V_{m\gamma}|^2}{(E_o + E_d)^2}\right]^{-1} . \tag{5.3.10}$$

The analysis of the behavior of $|S_{md}|^2$ as a function of E_m is performed in the previous section (its final part).

In the case when condition (5.3.9) is not satisfied, the exact energy levels of the system belong to the continuum

$$E \geq 0 . \tag{5.3.11}$$

In this case the unitary matrix elements $<m|E>$ have the form [130, 131, 132], (5.3.5),

$$|\langle m|E\rangle|^2=\frac{1}{\pi\rho(E)}\frac{\hbar\gamma(E)}{\hbar^2\gamma^2(E)+(\tilde{E}_m-E)^2}\ . \qquad (5.3.12)$$

Here $\rho(E)$ is the density of the states

$$dZ=\rho(E)dE\ , \qquad (5.3.13)$$

$$\tilde{E}_m(E)=E_m-\int_0^\infty\frac{|V_{m\gamma}|^2}{E_\gamma-E}\rho(E_\gamma)dE_\gamma\ , \qquad (5.3.14)$$

and

$$\gamma(E)=\frac{\pi}{\hbar}\sum_\gamma|V_{m\gamma}|^2\,\delta(E_\gamma-E)=\frac{\pi}{\hbar}\rho(E)|\overline{V_{m\gamma}}|^2\ , \qquad (5.3.15)$$

and $|\overline{V_{m\gamma}}|^2$ is some average perturbation energy matrix element, determined by Eq. (5.3.15).

This result shows that the configuration interaction "dilutes" the discrete state |m> throughout a band of exact stationary states whose profile is represented by a resonance curve with half-width $\pi\gamma(E)$. If the system under consideration were prepared in the |m> state, at a certain instant, it would autoionize with the mean lifetime γ^{-1} (to a certain ground state |g>).

5.4 Spontaneous emission of bosons (phonons) and tunneling in the rotating wave approximation

The problem of a two-state system is ubiquitous in physics and chemistry. In the simplest examples, the system possesses a degree of freedom that can take only two values, for example, the spin projection in the case of a nucleus of spin ½, the strangeness in the case of neutral K meson, or the polarization in the case of a photon. These are intrinsically two-state systems. A more common situation is when only two states of a multi-state system are relevant for certain problems. As has been shown (section 3.3) any two-state system can be described by the effective spin operators r_1, r_2 and r_3, satisfying usual commutation relations for spin ½.

In practice, almost every real-life two-state system interacts with its environment. In a quite general case, the electron-nuclear system is represented by two potential hypersurfaces – electronic terms, plus the perturbation causing transitions between these terms (3.3.6). These hypersurfaces are described in harmonic approximation by hyperparabolas, (3.3.9) and (3.3.10) (without anharmonic terms), with identical frequencies and different positions of minima. Such a model has been useful for the description of various processes, such as electronic transfer, energy transfer, tunneling in dissipative media, etc. (see section 3.5).

The Hamiltonian of the model, describing both nuclear and electronic motions, can be presented as [see (3.3.6), with $U_1 = J_1 + \frac{1}{2}\sum_k \omega_k^2 (q_k - q_{1k}^o)^2$ and $U_2 = J_2 + \frac{1}{2}\sum_k \omega_k^2 (q_k - q_{2k}^o)^2$]]

$$H = \frac{1}{2}\sum_k \left(p_k^2 + \omega_k^2 q_k^2\right) + n_1 E_1 + n_2 E_2 - n_1 \sum_k \omega_k^2 q_{1k}^o q_k - n_2 \sum_k \omega_k^2 q_{2k}^o q + 2r_1 V \; . \tag{5.4.1}$$

Here q_k and p_k are the operators of the coordinate and momentum of the k-th vibrational mode of the system, $n_1 = \frac{1}{2} + r_3$, $n_2 = \frac{1}{2} - r_3$; $n_1 = 1$, $n_2 = 0$ when $r_3 = \frac{1}{2}$, and $n_2 = 1$, $n_1 = 0$ when $r_3 = -\frac{1}{2}$, and $E_i = J_i + \frac{1}{2}\sum_k \omega_k^2 (q_{ik}^0)^2$ are energy constants. The perturbation energy matrix elements are, generally speaking, functions of q_k and p_k, and we consider the approximation of constant $V_{12} = V_{21}^* = V$ (Condon approximation).

The Hamiltonian (5.4.1) has become known in the literature as the "spin-boson" Hamiltonian. In this chapter we consider the symmetrical case (in case of the tunneling – symmetrical potential well) and present the spin-boson Hamiltonian in the form

$$H = \left(r_1 + \frac{1}{2}\right)\hbar\Delta_o + \sum_k a_k^+ a_k \hbar\omega_k + (r_+ + r_-)\sum_k B_k \left(a_k^+ + a_k\right) \; . \tag{5.4.2}$$

Here

$$\hbar\Delta_o = 2V; \qquad V = V_{12} = V_{21} \; , \tag{5.4.3}$$

$r_\pm = r_3 \pm i r_2$; a_k^+ and a_k are creation and annihilation operators of bosons (phonons) and

$$q_k^o = q_{1k}^o = -q_{2k}^o = B_k \sqrt{\frac{2}{\hbar\omega_k^3}} \; . \tag{5.4.4}$$

Parameters B_k describe the interaction between the tunneling system (section 3.5), or any other two-state system, and the dissipative system. In the case of the tunneling system, a localization in one of the potential wells is described by the operator r_3; the localization in the left well corresponds to $r_3 = \frac{1}{2}$, while $r_3 = -\frac{1}{2}$ corresponds to the localization in the right well.

The dynamics of a two-state system coupled to a dissipative environment, described by the Hamiltonian (5.4.2), is rather complex, as far as we try to overcome the approximations used in Chapter 3. This dynamics can be described using various approximations (see a comprehensive review of Leggett et al. [137] and a more traditional approach by Silbey and Harris [138]). On the other hand, another truncated rotating wave Hamiltonian is widely used

$$\tilde{H} = \left(r_1 + \frac{1}{2}\right)\hbar\Delta_o + \sum_k a_k^+ a_k \hbar\omega_k + \sum_k B_k\left(r_+ a_k + r_- a_k^+\right) \quad . \tag{5.4.5}$$

This Hamiltonian differs from the spin-boson Hamiltonian (5.4.2) by omitting the so-called "counter-rotating" terms r_+a_k and r_-a_k (rotating-wave approximation, RWA). It was originally suggested by Lee [139], intensively studied by Davidson and Kozak [140], and used by Pfeifer [141]. It should be emphasized that the Hamiltonian (5.4.5) by itself does not describe any specific physical system. It may provide only an approximate description of the time development of physical systems if the omitted counter-rotating terms can be considered as a small perturbation. This question will be considered later on.

The dynamics of the spin-boson system described by the truncated Hamiltonian (5.4.5) allows a consistent and exact analysis. In the context of the tunneling process, it has been shown [142, 143] that there exists a certain value for the matrix element of the perturbation energy between the states $r_3 = \pm\frac{1}{2}$ (left and right), $\hbar\Delta_o^{cr}$, such that for $|\Delta_o| > \Delta_o^{cr}$, quantum beats between two wells decay to zero at $t \to \infty$, while for matrix elements $|\Delta_o| < \Delta_o^{cr}$ the tunneling friction vanishes. This latter phenomenon has been called supertunneling. In the context of the relaxation of spin systems via spontaneous emission of phonons, it has been shown [131] that at the energy difference between two levels ($r_1 = \frac{1}{2}$ and $r_1 = -\frac{1}{2}$) of the spin system $|\Delta_o| < \Delta_o^{cr}$, there is a nonzero probability to remain in the excited state $r_1 = \frac{1}{2}$, $\Delta_o > 0$, at $t \to \infty$.

We now present exact solutions corresponding to the rotating wave Hamiltonian (5.4.5). We will use general relations of sections (5.1) and (5.2). The important quantity of the system is the energy shift function $R_a(E)$ of the unperturbed level a. First, let us consider the ground state of the unperturbed state $|-\frac{1}{2},0\rangle$. Here sign 0 means that the system is in the vacuum state, with all phonon (boson) numbers being zero. It is easy to check that the matrix elements of the perturbation energy, (5.4.5)

$$V = \sum_k B_k\left(r_+ a_k + r_- a_k^+\right) \quad , \tag{5.4.6}$$

are zero for all quantum numbers m in the vacuum state, $n_k = 0$

$$\left\langle -\frac{1}{2} 0 \,|\, V \,|\, m0 \right\rangle = 0 \quad . \tag{5.4.7}$$

This means that for $r_1 = -\frac{1}{2}$, $n_k = 0$

$$R_{-\frac{1}{2},0} = 0; \quad E_{-\frac{1}{2},0} = 0; \tag{5.4.8}$$

where $E_{-\frac{1}{2},0}$ is the exact energy level.

Thus the interaction (5.4.6) does not affect the ground state $|-\frac{1}{2},0>$, its energy remains the same and its weight equals 1; the system being initially in the state $|-\frac{1}{2},0>$ also remains in the same state.

Now let us consider the excited state $|\frac{1}{2},0>$. For this state, it follows from Eqs. (5.2.4), and (5.2.5), that

$$R_{\frac{1}{2},0}(E) = \sum_k \frac{|V_{\frac{1}{2}0;-\frac{1}{2},1_k}|^2}{E - \hbar\omega_k} , \tag{5.4.9}$$

$$E = \hbar\Delta_o + \sum_k \frac{|V_{\frac{1}{2}0;-\frac{1}{2},1_k}|^2}{E - \hbar\omega_k} , \tag{5.4.10}$$

$$|S_{\frac{1}{2},0;E}|^2 = \frac{1}{1 + \sum_k \frac{|V_{\frac{1}{2},0;-\frac{1}{2}k}|^2}{(E - \hbar\omega_k)^2}} . \tag{5.4.11}$$

These relations are isomorphous to those of the one-level system, interacting with the continuum states (5.3.3 – 5.3.5). The difference is in the existence of the ground state (5.4.8). The ground state is located in the outset of the continuum states of the phonon (boson) system. This ground state is also the exact ground state of the system. If the system is initially in the ground state, it remains unchanged.

We consider a spontaneous emission of bosons from the initially excited state. This means that at time $t = 0$, the spin is in the state $r_1 = \frac{1}{2}$, with energy

$$E_+ = E_{\frac{1}{2},0} = \hbar\Delta_o , \tag{5.4.12}$$

while phonons (bosons) are in the vacuum state

$$\Phi_{1...k...n} = |0_1...0_k...0_N> = \prod_k^N |\phi_0(q_k)> , \tag{5.4.13}$$

where $\phi_0(q_k)$ is an eigenstate of the k-th boson mode, and the index 0 means that the number of bosons in this state is zero. Thus we identify the state |m> of Sec. 3; it is that of the spin (two-level molecule) in its excited state $r_1 = \frac{1}{2}$ with no bosons (a vacuum state) in the dissipative state

$$|\frac{1}{2},0> = |\psi_{1/2}> \prod_k |\phi_0(q_k)> \quad , \tag{5.4.14}$$

its energy being (5.4.12).

This state is coupled by the interaction energy, (5.4.6), to a continuum of one-boson states

$$|\gamma> = |k> = |\psi_{-\frac{1}{2}}> \varphi_1(q_k)> \prod_{k'(\neq k)} |\varphi_0(q_{k'})> \quad , \tag{5.4.15}$$

with the energy eigenvalue

$$E_\gamma = E_k = \hbar\omega_k \quad . \tag{5.4.16}$$

It should be stressed that the interaction V, (5.4.6), connects the state |γ> with the discrete state $|\frac{1}{2},0>$ only. There are no matrix elements of the interaction V connecting |γ> with other continuous state $|\gamma'>$, as $|\frac{1}{2},k'>$. The full interaction energy (not a rotating wave one) containing $r_+ a_k^+$ and $r_- a_k$ terms, has such matrix elements. In this case the system is not isomorphous to the system with one discrete level (of the previous section).

Now, as in the previous section, we will consider two cases. The first case is when $\hbar\Delta_o$ is larger than the critical energy defined by the equation

$$\hbar\Delta_o > \hbar\Delta_c = \sum_k \frac{|B_k^2|}{\hbar\omega_k} \quad , \tag{5.4.17}$$

which follows from Eqs. (5.4.10) and (5.4.6). In this case the exact energies E form a continuous spectrum (without a discrete energy). Similarly to Eq. (5.3.12) we get

$$\left|S_{-\frac{1}{2},0;k}\right|^2 = \frac{1}{\pi\rho(\omega_k)} \frac{\gamma(\omega_k)}{\gamma^2(\omega_k) + (\Delta - \omega_k)^2} \quad . \tag{5.4.18}$$

Here $\rho(\omega_k)$ is a density of states k

$$dZ = \rho(\omega_k) d\omega_k \quad , \tag{5.4.19}$$

$$\Delta(\omega_k) = \Delta_0 - \frac{1}{\hbar^2}\int_0^\infty d\omega_{k'} \frac{|B_k|^2 \rho(\omega_{k'})}{\omega_{k'} - \omega_k} \equiv \Delta_0 - F(\omega_k) \quad , \tag{5.4.20}$$

and

$$\gamma(\omega_k) = \frac{\pi}{\hbar^2}\sum_{k'} |B_{k'}|^2 \delta(\omega_{k'} - \omega_k) = \frac{\pi}{\hbar^2}\rho(\omega_k)|\bar{B}_k|^2 \quad . \tag{5.4.21}$$

The time dependence of

$$P_+ = \rho_{\frac{1}{2}0;\frac{1}{2}0} \quad , \tag{5.4.22}$$

the probability to remain in the unperturbed state $|\frac{1}{2},0>$, is determined by Eq. (5.1.13) $(u = v = u' = v' = \frac{1}{2}, 0\ldots0_k\ldots)$ and Eq. (5.4.18)

$$P_+(t) = \frac{1}{\pi^2}\left|\int_0^\infty \frac{\gamma(\omega_k)e^{-i\omega_k t}d\omega_k}{\gamma^2(\omega_k) + (\Delta - \omega_k)^2}\right|^2 \quad . \tag{5.4.23}$$

It is easy to show that in the approximation in which one can neglect the frequency dependence of $\gamma(\omega_k)$ and $\Delta(\omega_k)$ in the vicinity of $\omega_k \approx \bar{\Delta}$,

$$\gamma(\omega_k) \approx \gamma(\bar{\Delta}); \qquad \Delta(\omega_k) \approx \bar{\Delta} \quad , \tag{5.4.24}$$

$$\Delta(\bar{\Delta}) - \bar{\Delta} = 0; \qquad \bar{\Delta} \approx \Delta_o + \frac{1}{\hbar^2}\sum_k \frac{|B_k^2|}{\Delta_o - \omega_k} \quad , \tag{5.4.25}$$

the time-dependence of $P_+(t)$ has the exponential form

$$P_+(t) = e^{-2\gamma t}; \qquad P_+(0) = 1 \quad . \tag{5.4.26}$$

This approximation coincides with the Markovian (section 2.4) (or Weisskopf-Wigner [144]) approximation. Exact time-dependence (5.4.23) has been analyzed by Davidson and Kozak [140].

In the case opposite to (5.4.17)

$$\Delta_o < \Delta_c = \sum_k \frac{|B_k|^2}{\hbar^2\omega_k} \quad , \tag{5.4.27}$$

there is the exact discrete level

$$E_d = -E_o; \qquad E_o > 0 \ , \tag{5.4.28}$$

which is the root of equation

$$E_o + \hbar\Delta_o = \sum_k \frac{|B_k^2|}{E_o + \hbar\omega_k} \ . \tag{5.4.29}$$

The exact level E_d=-E_o becomes the ground state, since $E_{-\frac{1}{2},0}$=0, according to Eq. (5.4.8). The weight of this exact discrete level, (5.4.11), is

$$\left|\left\langle\frac{1}{2},0|E_d\right\rangle\right|^2 = \left(1 + \sum_k \frac{|B_k|^2}{(E_o + \hbar\omega_k)^2}\right)^{-1} \ . \tag{5.4.30}$$

This means that according to Eq. (5.1.13), there is the nonvanishing probability that the system would remain in its initial state $|\frac{1}{2},0\rangle$, when $t \to \infty$

$$P_{\frac{1}{2}}(\infty) = P_+^{(\infty)} = \left(1 + \sum_k \frac{|B_k|^2}{(E_o + \hbar\omega_k)^2}\right)^{-2} \ . \tag{5.4.31}$$

It should be stressed that there is no positive discrete level. Such a discrete level would be degenerated with the energy continuum, and this means that there is nonvanishing $|S_{\frac{1}{2},0;E_d}|^2$, with $E_d > 0$. In this case the sum in the denominator of Eq. (5.4.11), when it is transformed into the integral, is diverging. This means that

$$|S_{\frac{1}{2},0;E_d}|^2 = 0; \qquad E_d > 0 \ . \tag{5.4.32}$$

The latter equation means that the exact discrete level, if this exists, should be nonpositive.

Equation (5.4.31) is characteristic for dissipationless behavior of the state with exact negative energy. Another phenomenon characterizing the dissipationless regime is the supertunneling [142,143,128] – nondecaying quantum beats between two potential (symmetrical) wells. The description of a tunneling in the terms of the effective spin is given in Section 3.5.

We are interested in the time development of the tunneling system interacting with the phonon bath. To describe the spontaneous emission, we needed the eigenfunctions of operator r_1. In the context of tunneling, these functions describe delocalized states, while the localization in one of the wells is described by the eigenfunctions of operator r_3

$$\psi_{L,R} = \frac{1}{\sqrt{2}}(\psi_{-\frac{1}{2}} \pm \psi_{\frac{1}{2}}) \ , \tag{5.4.33}$$

where L, R designate localization in the left and right wells, while $\psi_{\pm\frac{1}{2}}$ are eigenfunctions of operator r_1. The density matrix of the two-level system in the new (site) representation can be expressed through the density matrix in the r_1 representation

$$\sigma'_{mn} = \sum_{k\ell} U^{*}_{km}\sigma_{k\ell}U_{\ell n} \ , \tag{5.4.34}$$

where

$$\psi'_n = \sum_m U_{mn}\psi_m \ , \tag{5.4.35}$$

and

$$U_{-\frac{1}{2}L} = U_{\frac{1}{2}L} = U_{-\frac{1}{2}R} = -U_{\frac{1}{2}R} \ . \tag{5.4.36}$$

In particular, it follows from these relations that the probability to remain in the left well equals

$$P_L = \frac{1}{2} + \frac{1}{2}(\sigma_{-\frac{1}{2},\frac{1}{2}} + \sigma_{\frac{1}{2},-\frac{1}{2}}) \ . \tag{5.4.37}$$

We assume that initially (t = 0) the system was in the left well

$$\sigma_{\frac{1}{2},\frac{1}{2}}(0) = \sigma_{-\frac{1}{2},-\frac{1}{2}}(0) = \sigma_{-\frac{1}{2},\frac{1}{2}}(0) = \sigma_{\frac{1}{2},-\frac{1}{2}}(0) = \frac{1}{2} \ . \tag{5.4.38}$$

It should be noticed that we consider the zero temperature of the phonon bath.

The derivation similar to that of spontaneous emission leads to the following description of the tunneling

$$P_L = \begin{cases} \frac{1}{2}[1 + e^{-\gamma t}\cos\frac{\Delta_o}{\hbar}t], & \Delta_o > \Delta_c = \sum_k \frac{B_k^2}{\hbar^2\omega_k} \\ \frac{1}{2}[1 + \cos\frac{\Delta_o}{\hbar}t], & \Delta_o < \Delta_c \end{cases} \ , \tag{5.4.39}$$

where γ is determined by Eqs. (5.4.21) and (5.4.24). The Markovian approximation has been used for the description of the dissipation regime of tunneling (the first of the expressions in the r.h.s. of Eq. (5.4.39)). The dissipationless regime at $\Delta_o < \Delta_c$ is characterized by the nondecaying quantum beats in the tunneling system coupled to the phonon bath, at zero temperature, T = 0.

Apart from the emerging "new" bound states in the configurational interaction, "frozen" upper states in the spontaneous emission, and nondecaying tunneling quantum beats, there are other processes having the same physics and mathematics.

The ionization can be caused by the electromagnetic field whose frequency ω is larger than $I_o/\hbar$, where I_o is the ionization energy. This energy equals the energy difference between the onset of the continuum states and ground state energy, E_g. In a sense, the electromagnetic field of frequency $\omega > \frac{I_0}{\hbar}$ causes the appearance of a new discrete state $\tilde{E}_g$ degenerate with the continuum

$$\tilde{E}_g = E_g + \hbar\omega \quad . \tag{5.4.40}$$

This situation is similar to that described by the configuration interaction with one discrete level. The interaction with the electromagnetic field may "push" the discrete level $\tilde{E}_g$ below the continuum of states. In this case the ionization is practically stopped (provided the weight of the new discrete level is high enough). It may be said that the interaction with the electromagnetic field creates an "artificial" autoionizing state, which leads to trapping of electronic population in the ground state when condition (5.3.9) of the emerging of the exact discrete states is satisfied. This situation has been explored in a number of works [145-148]. The situation close to that considered in this section has been considered by Kofman et al. [149], analyzing spontaneous emission in photonic band structures [150, 151].

5.5 Weak coupling case (beyond RWA)

We will now consider the full Hamiltonian (5.4.2) containing counterrotating terms $r_+a_k^+$ and r_-a_k as well. The spin system itself is described by the first term in Eq. (5.4.2). It has two levels, corresponding to two eigenvalues of operator $r_1 = \pm\frac{1}{2}$. Considering both positive and negative values of Δ_o, we can use the first term in a symmetric way

$$E_{spin} = \left(r_1 + \frac{1}{2}\frac{|\Delta_o|}{\Delta_o} \right) \hbar\Delta_o \quad . \tag{5.5.1}$$

This spin Hamiltonian has zero ground state energy, and the upper state energy is $\hbar|\Delta_o|$. For

$$\Delta_o > 0, \qquad E_{\frac{1}{2}} = E_+ = \hbar\Delta_o, \qquad E_{-\frac{1}{2}} = E_- = 0 \ , \tag{5.5.2}$$

while for

$$\Delta_o < 0, \qquad E_{-\frac{1}{2}} = E_+ = -\hbar\Delta_o, \qquad E_{\frac{1}{2}} = E_- = 0 \ . \tag{5.5.3}$$

We continue to designate spin values $\pm\frac{1}{2}$, only for the eigenvalues of spin component r_1, while for the eigenvalues of spin components r_3 we use connotations L (left) for $r_3 = \frac{1}{2}$, and R (right) for $r_3 = -\frac{1}{2}$. We will be interested in the dissipationless regimes of the spin-boson system described by the Hamiltonian (5.4.2).

Two functions characterize the dissipationless regimes. First, this is

$$P_+^\infty(\Delta_o) = \lim_{t\to\infty} P_+(\Delta_o, t) \ , \tag{5.5.4}$$

where $P_+(\Delta_o,t)$ is the probability to remain in the upper level E_+, at time t, if initially $P_+(\Delta_o,0) \neq 0$. This probability depends on Δ_o, and our task is to analyze $P_\pm^\infty(\Delta_o)$ as functions of Δ_o. In the dissipationless regime $P_+^\infty \neq 0$, while in the dissipation regime the probability to remain on the upper level E_+ tends to zero, when $t \to \infty$.

Another dissipationless phenomenon is supertunneling [142,143,128] – nondecaying quantum beats between the symmetric potential wells. This phenomenon is characterized by the probability to remain in the left well (5.4.37)

$$P_L(t) = \frac{1}{2} + \frac{1}{2}(\rho_{-\frac{1}{2},\frac{1}{2}}(t) + \rho_{\frac{1}{2},-\frac{1}{2}}(t)) \ . \tag{5.5.5}$$

If $\rho_{mn}(0) = \frac{1}{2}$, for all (m,n) eigenvalues of $r_1 = \pm\frac{1}{2}$, then the initial probabilities are $P_L(0) = 1$; $P_R(0) = 0$. In the dissipation regime, quantum beats are characterized by functions

$$\rho_{-\frac{1}{2},\frac{1}{2}}(t) = \rho^*_{\frac{1}{2},-\frac{1}{2}}(t) \ , \tag{5.5.6}$$

which tend to zero when $t \to \infty$. On the other hand, in the dissipationless regime the quantum beats do not decay, and amplitudes of quantum beats

$$\lim_{t\to\infty} \rho_{-\frac{1}{2},\frac{1}{2}}(\Delta_o, t) = \rho^\infty_{-\frac{1}{2},\frac{1}{2}}(\Delta_o) \neq 0 \ , \tag{5.5.7}$$

may serve as a characteristic of the dissipationless regime.

We will analyze $P_+^\infty(\Delta_o)$ and $\rho^\infty_{-\frac{1}{2},\frac{1}{2}}(\Delta_o)$ as functions of Δ_o. First, we will show that there exists a certain range of the spin-boson Hamiltonian parameters which corresponds to the dissipationless regime. We will employ results of Section 5.2 and consider case (a) of this section: $\lim_{E_o \to 0} R_m(E_o)$ exists. Then, according to the general theory there is a critical frequency

$$\Delta_c = -\frac{1}{\hbar} R_{\frac{1}{2}}(0) \ , \tag{5.5.8}$$

such that in the region

$$-\Delta_c < \Delta_o < \Delta_c \ , \tag{5.5.9}$$

the spin-boson system reveals the nondissipative behavior. It is worthwhile to emphasize that this conclusion about the existence of the nondissipative regime is quite general. It is not connected with any specific approximation. Of course, to calculate Δ_c we need the approximation. In the general case, according to Eqs. (5.2.8) and (5.4.2), the critical frequency Δ_c satisfies the inequality

$$\Delta_c \geq \sum_k \frac{B_k^2}{\hbar^2 \omega_k} \ . \tag{5.5.10}$$

For the rotating wave Hamiltonian the critical frequency obtains its minimal value, (5.4.17)

$$\Delta_c = \sum_k \frac{B_k^2}{\hbar^2 \omega_k} \ . \tag{5.5.11}$$

Our task is to describe quantities $P_+^\infty(\Delta_o)$ and $\rho^\infty_{-\frac{1}{2},\frac{1}{2}}(\Delta_o)$ as functions of Δ_o in the dissipationless region (5.5.9); outside this region these two functions are simply zero. In this section we consider the weak coupling case

$$\sum\nolimits_\infty = 2\sum_k \frac{B_k^2}{\hbar^2 \omega_k^2} << 1 \ . \tag{5.5.12}$$

Here $\sum_\infty$ is a dimensionless coupling parameter, while the use of the connotation ∞ will be clear from the following discussion. In the weak coupling case the higher order terms, in the energy shift function $R_a(E_L)$ (5.2.5) can be neglected [128, 142]. In this case the critical frequency Δ_c is equal to (5.5.11). It should be mentioned that for the full spin-boson Hamiltonian (5.4.2), Δ_c, presented by Eq. (5.5.11), is an approximate value of this quantity, while for the truncated, rotating wave Hamiltonian (5.4.5), formula (5.5.11) is exact.

To proceed further, we have to assume a certain frequency dependence of the coupling parameter $B_k(\omega_k)$. Having in mind that the summation $\sum_k$ in the three-dimensional manifold can be reduced to integration $\int_0 \omega^2 d\omega$, we define

$$\sum_k B_k^2 \to \int_0 d\omega\omega^2 \, B^2(\omega) \ . \tag{5.5.13}$$

The frequency dependence of $B^2(\omega)$ may be taken in the form

$$B^2(\omega) = \frac{\hbar^2 \gamma_s}{\omega_c^2} \left(\frac{\omega}{\omega_c} \right)^{s-2} \exp(-\omega/\omega_c) \ , \tag{5.5.14}$$

where γ_s is a constant having the dimensionality of frequency.

This form of the $B(\omega)$ frequency dependence corresponds to the notation of Leggett et al. [137] for the spectral density function

$$J(\omega) = A_s \omega^s \exp(-\omega/\omega_c) \ . \tag{5.5.15}$$

Here ω_c is the cutoff frequency, which is assumed to be much larger than Δ_0. The integration in relation (5.5.13) is then extended to infinity. Another possibility which we will use, is to extend the integration to ω_D - the Debye frequency. In this case we do not use the exponent $\exp(-\omega/\omega_c)$ and put $\omega_c = \omega_D$.

The case s = 3 corresponds to the coupling constant and density of states for three-dimensional phonons interacting with a localized system. The case s = 1 is the ohmic case, and it also corresponds to the interaction with the acoustic phonons used in physisorption kinetics [153, 131].

In the weak coupling approximation (5.5.12), the equation determining the exact discrete excited state $E_d = -E_0$, $E_0 > 0$, can be found from Eq. (5.2.7)

$$\hbar\Delta_0 + E_0 = \sum_k \frac{B_k^2}{E_0 + \hbar\omega_k} = -R_{1/2}(-E_0) = -R_{1/2}(E_d) \ , \tag{5.5.16}$$

where index $\frac{1}{2}$ designates spin $r_1 = \frac{1}{2}$ (an excited state for $\Delta_0 > 0$). The coefficient $S_{1/2,0;d}$ of the unitary transformation from the state $|\frac{1}{2},0>$ (zero sign means the vacuum state, all $n_k = 0$) of the unperturbed Hamiltonian

$$H_0 = \left(r_1 + \frac{1}{2} \right) \hbar\Delta_0 + \sum_k a_k^+ a_k \hbar\omega_k \ , \tag{5.5.17}$$

to the discrete state of the exact Hamiltonian (5.4.2), has the form (5.2.6)

$$|S_{1/2,0;d}|^2 = \left|\left\langle \frac{1}{2},0 \,|\, d \right\rangle\right|^2 = \frac{1}{1-R'_{1/2}(E_d)} = \frac{1}{1+\sum_k B_k^2/(E_0+\hbar\omega_k)^2} \quad . \tag{5.5.18}$$

Function $-R_{1/2}(-E_0)$ can be calculated using relations (5.5.13) and (5.5.14) with a cutoff frequency ω_D (the Debye model)

$$-R_{1/2}(-E_0) = \sum_k \frac{B_k^2}{E_0+\hbar\omega_k} = \frac{\hbar^2\gamma_s}{\omega_D^s}\int_0^{\omega_D} \frac{\omega^s}{E_0+\hbar\omega}d\omega$$
$$= \frac{\hbar^2\gamma_s}{\omega_D^s}\left[\frac{\omega_D^s}{s\hbar} - \frac{E_0\omega_D^{s-1}}{(s-1)\hbar^2} + \frac{E_0^2\omega_D^{s-2}}{(s-2)\hbar^3}\right.$$
$$\left. \ldots + (-1)^{s-1}\frac{E_0^{s-1}\omega_D}{\hbar^s} + \frac{(-1)^s E_0^s}{\hbar^{s+1}}\ln\left(1+\frac{\hbar\omega_D}{E_0}\right)\right] \ , \tag{5.5.19}$$

where $s \geq 1$ (s is an integer number – see Ref. [154]).

In a quite good approximation we can assume

$$E_0 << \omega_D \ . \tag{5.5.20}$$

Then for all values of E_0

$$-R_{1/2}(-E_0) \approx -R_{1/2}(0) = \sum_k \frac{B_k^2}{\hbar\omega_k} = \hbar\Delta_c = \hbar\gamma_s/s \ , \tag{5.5.21}$$

and

$$-R'_{1/2}(0) = \sum_k \frac{B_k^2}{\hbar^2\omega_k^2} = \frac{1}{2}\sum\nolimits_\infty = \frac{\gamma_s}{\omega_D(s-1)} \quad (\text{for } s>1) \ , \tag{5.5.22}$$

$$-R'_{1/2}(E_d) = \ln\left(1+\frac{\hbar\omega_D}{E_0}\right) \approx \ln\frac{\hbar\omega_D}{E_0} \quad (\text{for } s=1) \ . \tag{5.5.23}$$

In the ohmic case ($s = 1$) the coupling parameter $\frac{1}{2}\sum_\infty = -R'_{1/2}(0)$ diverges

$$\lim_{E_0\to 0} R'_{1/2}(E_0) = \infty \ . \tag{5.5.24}$$

Therefore, one cannot treat the ohmic case in the weak coupling approximation (5.5.12). On the other hand, the cases $s > 1$ can be treated in the weak coupling approximation (5.5.12), provided that

$$\frac{\gamma_s}{\omega_D} = \frac{s\Delta_c}{\omega_D} << 1 \quad \text{and} \quad E_0 << \omega_D \ . \tag{5.5.25}$$

In this case, we get from Eqs. (5.5.18), (5.5.12) and (5.5.22)

$$\left|S_{1/2,0;d}\right|^2 \approx 1 - \frac{1}{2}\sum\nolimits_\infty \quad , \tag{5.5.26}$$

and

$$-E_d = E_0 \approx \hbar(\Delta_c - |\,\Delta_0\,|)\, ; \ \Delta_c = \gamma_s/s \ . \tag{5.5.27}$$

Similarly for the ground state |g> we get

$$-R_{-\frac{1}{2}}(E_g) = E_g = -\sum_k \frac{B_k^2}{\hbar\Delta_0 + \hbar\Delta_k - E_g} \approx -\hbar\Delta_c \quad , \tag{5.5.28}$$

so that the energy difference between the exact excited and ground states equals

$$E_d - E_g = \hbar\Delta_0 \quad , \tag{5.5.29}$$

while the unitary matrix element $|\,S_{-\frac{1}{2},0;g}\,|^2$ equals

$$|\,S_{-\frac{1}{2},0;g}\,|^2 = 1 - \sum_k \frac{B_k^2}{(\hbar\Delta_0 + \hbar\omega_k - E_g)^2} \approx 1 - \sum_k \frac{B_k^2}{\hbar^2\omega_k^2} = 1 - \frac{1}{2}\sum\nolimits_\infty \ . \tag{5.5.30}$$

Now let us assume that initially the system is in the state characterized by density matrix elements (the boson field is in its vacuum state: $n_k = 0$)

$$\rho_{mn}(0); \qquad m,n = \pm\frac{1}{2} \ , \tag{5.5.31}$$

where m,n are eigenvalues of the operator r_1. Using Eq. (5.1.13), expressions for the unitary matrix elements $S_{1/2,0;d}$, (5.5.26), $S_{-1/2,0,g}$, (5.5.30), and others such as $S_{-1/2,1_\lambda;d}$, $S_{1/2,1_\lambda;g}$, etc. (the derivation of these matrix elements is skipped here), the following asymptotic equations may be derived

$$\rho^\infty_{\frac{1}{2},\frac{1}{2}} = \rho_{\frac{1}{2},\frac{1}{2}}(0)(1 - \sum\nolimits_\infty) + \rho_{-\frac{1}{2},-\frac{1}{2}}(0)\sum\nolimits_\infty \quad , \tag{5.5.32}$$

$$\rho^\infty_{-\frac{1}{2},-\frac{1}{2}} = \rho_{\frac{1}{2},\frac{1}{2}}(0)\sum\nolimits_\infty + \rho_{-\frac{1}{2},-\frac{1}{2}}(0)(1 - \sum\nolimits_\infty) \ , \tag{5.5.33}$$

$$\rho^{\infty}_{-\frac{1}{2},\frac{1}{2}} = \left(\rho^{\infty}_{\frac{1}{2},-\frac{1}{2}}\right)^{*} = \rho_{\frac{1}{2},-\frac{1}{2}}(0)\sum\nolimits_{\infty} + \rho_{-\frac{1}{2},\frac{1}{2}}(0)(1-\sum\nolimits_{\infty}) \ , \qquad (5.5.34)$$

where $\rho^{\infty}_{-\frac{1}{2},\frac{1}{2}}$ and $\rho^{\infty}_{\frac{1}{2},-\frac{1}{2}}$ are the amplitudes of the exponentials $e^{-i\Delta_0 t}$ and $e^{i\Delta_0 t}$ respectively.

Matrix elements ρ^{∞}_{mn} do not depend on Δ_0 in the whole dissipationless region (5.5.9) (see (5.5.21), while outside this region $\rho^{\infty}_{\frac{1}{2},\frac{1}{2}} = \rho^{\infty}_{-\frac{1}{2},\frac{1}{2}} = 0$ (for $\Delta_0 > \Delta_c$). In deriving the above equations we did not use the rotating wave approximation. In the latter approximation (section 5.4) the ground state $|-\frac{1}{2}>$ is not affected by the interaction (5.4.6): $R_{-\frac{1}{2},0} = 0;\ E_g = 0;\ |S_{-\frac{1}{2}0;g}|^2 = 1$, (5.4.8), while in the weak coupling approximation these quantities are given by Eqs. (5.5.28) nd (5.5.30).

5.6 Semiquantitive analysis of the dissipationless regime

As has been mentioned above, our task is to describe asymptotic ($t \to \infty$) quantities $P^{\infty}_{+}(\Delta_0)$ and $\rho^{\infty}_{-\frac{1}{2},\frac{1}{2}}(\Delta_0)$ as functions of Δ_0. We have fulfilled this task in the weak coupling case of small $\sum_{\infty} << 1$. In the general case of arbitrary Σ_{∞} we are able to describe functions $P^{\infty}_{+}(\Delta_0)$ and $\rho^{\infty}_{-\frac{1}{2},\frac{1}{2}}(\Delta_0)$ in a semiquantitive way. According to the analysis performed in the last part of Section 5.2, the exact discrete energy $E_d(\Delta_0)$ declines monotonically when Δ_0 changes from 0 to Δ_c, while $|S_{md}|^2 = |<m|d>|^2$ decreases from its maximum value at $\Delta_0 = 0$ to zero at $\Delta_0 = \Delta_c$.

However, quantity $P^{\infty}_{+}(\Delta_0)$ - the probability to remain on the upper level, when $t \to \infty$, does not coincide with $|<\frac{1}{2},0|E_d>|^2$ in the general case. But it can be shown that $P^{\infty}_{+}(\Delta_0)$ is proportional to quantities $|<\frac{1}{2},0|E_d>|^2$. Let us assume that initially the two-level system is at the upper level $E_+ = \hbar\Delta_0 > 0$. Then the asymptotic probability ($t \to \infty$) to remain on this level can be derived from the Hamiltonian (5.4.2) and Eqs. (5.1.13) and (5.1.16)

$$P^{\infty}_{+} = P^{\infty}_{\frac{1}{2},0} + \sum_{\lambda\lambda'} P^{\infty}_{\frac{1}{2},1_\lambda,1_{\lambda'}} + \sum_{\lambda,\lambda',\lambda'',\lambda'''} P^{\infty}_{\frac{1}{2},1_\lambda,1_{\lambda'},1_{\lambda''},1_{\lambda'''}} + \ldots$$

$$= S_{\frac{1}{2}0;d} S^{*}_{\frac{1}{2}0;d} \rho_{\frac{1}{2}0;\frac{1}{2}0}(0) S_{\frac{1}{2}0;d} S_{\frac{1}{2},0d} + \sum_{\lambda\lambda'} S_{\frac{1}{2},1_\lambda,1_{\lambda'};d} S^{*}_{\frac{1}{2}0;d} \rho_{\frac{1}{2}0,\frac{1}{2}0}(0) S_{\frac{1}{2}0;d} S^{*}_{\frac{1}{2}1_\lambda,1_{\lambda'};d} + \ldots$$

$$=|S_{\frac{1}{2}0;d}|^2\,(|S_{\frac{1}{2}0;d}|^2+\sum_{\lambda\lambda'}|S_{\frac{1}{2}1_\lambda,1_{\lambda'};d}|^2+\ldots)\,; \qquad \rho_{\frac{1}{2}0;\frac{1}{2}}(0)=1 \ . \tag{5.6.1}$$

In the general case

$$P_+^\infty(\Delta_0)=|S_{\frac{1}{2}0;d}|^2\left(|S_{\frac{1}{2}0;d}|^2+\sum_{\lambda\lambda'}|S_{\frac{1}{2}1_\lambda 1_\lambda;d}|^2+\ldots\right)\rho_{\frac{1}{2}0;\frac{1}{2}0}(0)$$

$$+|S_{-\frac{1}{2}0;g}|^2\left(\sum_\lambda|S_{\frac{1}{2}1_\lambda;g}|^2+\sum_{\lambda,\lambda',\lambda''}|S_{\frac{1}{2}1_\lambda,1_{\lambda'},1_{\lambda''};g}|^2+\ldots\right)\rho_{-\frac{1}{2}0;-\frac{1}{2}0}(0) \ . \tag{5.6.2}$$

In order to describe semiquantitatively $P_+^\infty(\Delta_0)$ (and $\rho^\infty_{\frac{1}{2},\frac{1}{2}}(\Delta_0)$) as a function of Δ_0, we need an exact solution for matrix elements $\rho^\infty_{\frac{1}{2},\frac{1}{2}}$ and $\rho^\infty_{-\frac{1}{2},-\frac{1}{2}}$ at $\Delta_0 = 0$. The exact solution for matrix elements $\rho_{\frac{1}{2},\frac{1}{2}}(t)$ and $\rho_{-\frac{1}{2},-\frac{1}{2}}(t)$ has been found in the case $\Delta_0 = 0$ and $\rho_{\frac{1}{2}0;\frac{1}{2}0}(0)=1$ [155]. (This solution cannot be applied to an interaction with photons, due to the existence of the last term in the r.h.s. of Eq. (4.1.7).) These results can be generalized for the initial (vacuum) condition [128]

$$\rho_{mn}(0)=\rho_{m,0;n,0}(0), \qquad m,n=\pm\frac{1}{2} \ , \tag{5.6.3}$$

$$\rho_{\frac{1}{2},\frac{1}{2}}(t)=\rho_{\frac{1}{2},\frac{1}{2}}(0)e^{-\Sigma(t)}\cosh\Sigma(t)+\rho_{-\frac{1}{2},-\frac{1}{2}}(0)e^{-\Sigma(t)}\sinh\Sigma(t) \ , \tag{5.6.4}$$

$$\rho_{-\frac{1}{2},-\frac{1}{2}}(t)=\rho_{\frac{1}{2},\frac{1}{2}}(0)e^{-\Sigma(t)}\sinh\Sigma(t)+\rho_{-\frac{1}{2},-\frac{1}{2}}(0)e^{-\Sigma(t)}\cosh\Sigma(t) \ , \tag{5.6.5}$$

$$\rho_{\frac{1}{2},-\frac{1}{2}}(t)=\rho^*_{-\frac{1}{2},\frac{1}{2}}(t)=\rho_{\frac{1}{2},-\frac{1}{2}}(0)e^{-\Sigma(t)}\cosh\Sigma(t)+\rho_{-\frac{1}{2},\frac{1}{2}}(0)e^{-\Sigma(t)}\sinh\Sigma(t) \ , \tag{5.6.6}$$

where

$$\Sigma(t)=2\sum_k\frac{B_k^2}{\hbar^2\omega_k^2}[1-\cos\omega_k t] \ . \tag{5.6.7}$$

Using relations (5.5.13) and (5.5.14) we can present expression (5.6.7) as the integral

$$\Sigma(t)=\frac{2\gamma_s}{\omega_c^s}\int_0^\infty d\omega\,\omega^{s-2}e^{-\omega/\omega_c}(1-\cos\omega t) \ . \tag{5.6.8}$$

For s (integer) and higher than 1 the integral $\Sigma(t)$ has finite asymptotic ($t \to \infty$) value [154]

$$\sum_{\infty} = 2(s-2)!\frac{\gamma_s}{\omega_c} \quad (s \geq 2) \ . \tag{5.6.9}$$

In the ohmic case (s = 1) the integral may be calculated as [154]

$$\sum(t) = \frac{\gamma_1}{\omega_c}\ln\left(1+\omega_c^2 t^2\right) \ . \tag{5.6.10}$$

This integral diverges

$$\sum_{\infty} = \lim_{t\to\infty}\sum(t) = \infty \ . \tag{5.6.11}$$

This conclusion coincides with relation (5.5.24) obtained in the Debye model with a finite upper limit ω_D of integral (5.5.19) (without the cutoff factor $\exp(-\omega/\omega_c)$). The nonessential difference between $\sum_{\infty}$ of Eq. (5.6.9) and Eq. (5.5.22) is connected with the usage of different cutoff models (ω_D and ω_c).

Now, we can summarize the description of the nondissipative regime in the spin-boson system. The existence of the nondissipative regime, for $s > 1$, has been proven quite generally. This result is not connected with the use of weak or strong coupling. Relation (5.5.10) for Δ_c determines the range of the dissipationless regime (everything is considered at zero absolute temperature $T = 0$).

In the range of parameters

$$-\Delta_c < \Delta_0 < \Delta_c \ , \tag{5.6.12}$$

spin-boson systems reveal the dissipationless behavior. As has been mentioned above, two phenomena characterize the dissipationless regime. Considering two levels of the spin system, the upper one E_+ and the lower E_-, we can define the asymptotic value ($t \to \infty$) of the probability to remain in the excited state P_+^{∞}. While in the dissipative region ($|\Delta_0|>\Delta_c$) P_+^{∞} is zero, in the dissipationless regime there exists nonzero probability to remain in the excited state, $P_+^{\infty} \neq 0$. At $\Delta_0 = 0$, and in its vicinity, we can find P_+^{∞} using exact (for $\Delta_0 = 0$) Eqs. (5.6.4). Since initial ($t = 0$) probabilities $|c_{1/2}|^2 = P_+(0)$ for positive Δ_0, and $|c_{-\frac{1}{2}}|^2 = P_+(0)$ for negative Δ_0, we can obtain from Eqs. (5.6.4 and 5.6.5)

$$P_+^{\infty} = P_+(0)e^{-\sum_{\infty}}\cosh\sum_{\infty} + P_-(0)e^{-\sum_{\infty}}\sinh\sum_{\infty},$$

$$P_-^\infty = P_+(0)e^{-\Sigma_\infty}\sinh\Sigma_\infty + P_-(0)e^{-\Sigma_\infty}\cosh\Sigma_\infty \quad . \tag{5.6.13}$$

Plots of P_+^∞ and P_-^∞ as functions of Δ_0 for various initial values are shown in Fig. 6. Fig. 6a corresponds to the weak coupling case $\Sigma_\infty << 1$ (5.5.32, 5.5.33); as is shown in this plot $P_\pm^\infty$ depends on the initial values $P_\pm(0)$(nonergodic behavior). Fig. 6b describes the strong coupling case

$$\Sigma_\infty >> 1 \quad . \tag{5.6.14}$$

In the case of large Σ_∞ the asymptotic values $P_\pm^\infty$ take the form

$$P_+^\infty = P_-^\infty = \frac{1}{2} \quad . \tag{5.6.15}$$

In the case of weak coupling we can obtain the analytical solution for $P_\pm^\infty$ in the whole region (5.5.32-5.5.34), and determine Δ_c, Eq. (5.5.21). In the general and strong coupling cases we can obtain values of $P_\pm^\infty$ in the vicinity of $\Delta_0 = 0$, and lower limit of the critical frequency Δ_c (5.5.10). Thus, in the general case we can perform the semiquantitive analysis of the dissipationless regime (for arbitrary strength of coupling).

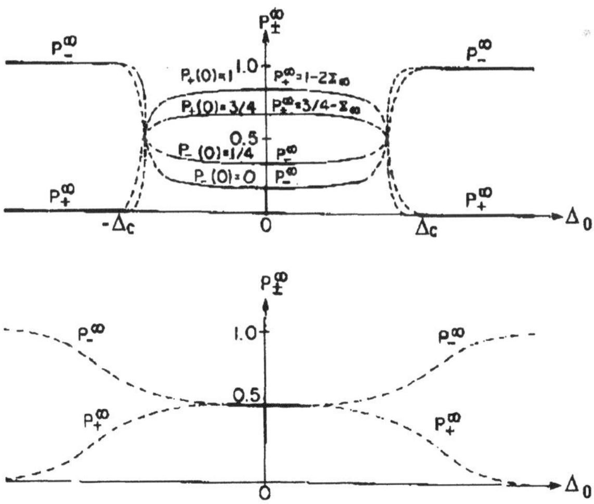

Fig. 6. (a) Weak coupling: $\Sigma_\infty << 1$. Functions $P_\pm^\infty$ are depicted at various initial conditions $P_+(0)$. (b) Strong coupling $\Sigma_\infty >> 1$. Points $\pm\Delta_c$ are not shown since they may exceed the maximum frequency of Δ_0. Exact solution is known at point $\Delta_0 = 0$.

$$P_L = \frac{1}{2} + \frac{1}{2}\left[\rho_{1/2,-1/2}(t) + \rho_{-1/2,1/1}(t)\right] , \tag{5.6.16}$$

where $\pm\frac{1}{2}$ are the eigenvalues of spin component r_1. In the dissipative regime the second term in the right-hand side of Eq. (5.6.16) describes decaying quantum beats so that the asymptotic values of the matrix elements vanish

$$\rho^{\infty}_{1/2,-1/2} = \rho^{\infty}_{-1/2,1/2} = 0 \ ,$$

and the system can be found with equal probabilities in both wells

$$P^{\infty}_L = P^{\infty}_R = \frac{1}{2} \ . \tag{5.6.17}$$

In the dissipationless regime, in the weak coupling approximation the asymptotic values of the density matrix elements in Eq. (5.6.16) are given by Eq. (5.5.34)

$$\rho^{\infty}_{\frac{1}{2},-\frac{1}{2}} e^{-i\Delta_0 t} ; \qquad \rho^{\infty}_{-\frac{1}{2},\frac{1}{2}} e^{i\Delta_0 t} \ , \tag{5.6.18}$$

i.e. quantum beats do not decay.

In the general case, we know the exact density matrix elements $\rho_{1/2,-1/2}(t)$ = $\rho^{*}_{-1/2,1/2}(t)$ (both strong and weak coupling) for $\Delta_0 = 0$, Eqs. (5.6.6) and (5.6.7). In a good approximation, solution (5.6.6) and (5.6.7) can be used in the vicinity of $\Delta_0 = 0$ (for $\Delta_0 << \Delta_c$). In this case $\Delta_0 = 0$ the spin-boson system is degenerate and this degeneracy can be removed by using the correct zeroth-approximation eigenfunction [142], so that in the first order approximation of the perturbation energy $(r_1 + \frac{1}{2})\hbar\Delta_0$ in the Hamiltonian (5.4.2), one gets the "dressed" quantum beats frequency

$$\Delta = \Delta_0 \exp\left[-2\sum_k \frac{B_k^2}{\hbar^2\omega_k^2}(2n_k + 1)\right] . \tag{5.6.19}$$

For the case of the zero-temperature boson system ($n_k = 0$) the "dressed" frequency takes the form

$$\Delta = \Delta_0 \exp[-\sum\nolimits_{\infty}] \ , \tag{5.6.20}$$

and the asymptotic probability to remain in the left (right) well has the form

$$P_{L,R} = \frac{1}{2} \pm \frac{1}{2}\left(\rho^{\infty}_{-1/2,1/2} e^{i\Delta t} + \rho^{\infty}_{1/2,-1/2} e^{-i\Delta t}\right) . \tag{5.6.21}$$

Here the asymptotic density matrix elements (for s > 1) could be derived from Eq. (5.6.6)

$$\rho^{\infty}_{1/2,-1/2} = (\rho^{\infty}_{-1/2,1/2})^{*} = \rho_{-\frac{1}{2},\frac{1}{2}}(0)e^{-\sum_{\infty}} \sinh \sum_{\infty} + \rho_{\frac{1}{2},-\frac{1}{2}}(0)e^{-\sum_{\infty}} \cosh \sum_{\infty}. \quad (5.6.22)$$

In the strong coupling case $\sum_{\infty} >> 1$

$$\rho^{\infty}_{-1/2,1/2} = -\frac{1}{2}\left(\rho_{\frac{1}{2},-\frac{1}{2}}(0) + \rho_{-\frac{1}{2},\frac{1}{2}}(0)\right). \quad (5.6.23)$$

In the ohmic case (s = 1) there are no quantum beats since, $\sum_{\infty} = \infty$ (5.6.1), and $\Delta = 0$.

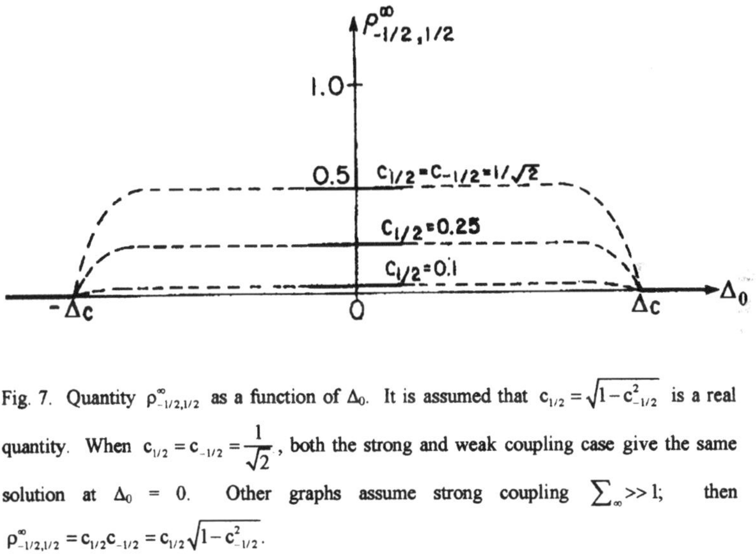

Fig. 7. Quantity $\rho^{\infty}_{-1/2,1/2}$ as a function of Δ_0. It is assumed that $c_{1/2} = \sqrt{1-c^2_{-1/2}}$ is a real quantity. When $c_{1/2} = c_{-1/2} = \frac{1}{\sqrt{2}}$, both the strong and weak coupling case give the same solution at $\Delta_0 = 0$. Other graphs assume strong coupling $\sum_{\infty} >> 1$; then $\rho^{\infty}_{-1/2,1/2} = c_{1/2}c_{-1/2} = c_{1/2}\sqrt{1-c^2_{-1/2}}$.

In Fig. 7 we plot $\rho^{\infty}_{-1/2,1/2}$ for $\rho_{\frac{1}{2},-\frac{1}{2}}(0) = c_{\frac{1}{2}}c_{-\frac{1}{2}}$ and for various (real) $c_{1/2} = \sqrt{1-c^2_{-1/2}}$. Both Figs. 6 and 7 present the results of the semiquantitive analysis. Exact solutions for $P^{\infty}_{\pm}$ and $\rho^{\infty}_{-1/2,1/2}$ are known at $\Delta_0 = 0$, while for $\Delta_0 << \Delta_c$ one can use these solutions as approximate ones. In the whole region of the dissipationless regime ($|\Delta_0| < \Delta_c$) the qualitative behavior of quantities $P^{\infty}_{\pm}$ can be deduced from the general conclusions of Section 5.2 (see the paragraph below Eq. (5.2.9) and Eqs. (5.6.1) and (5.6.2)). Similar conclusions can be obtained for quantity $\rho^{\infty}_{-\frac{1}{2},\frac{1}{2}}$.

5.7 Rotating wave, Markovian, and weak coupling approximations

There are two popular approximations in the theory of relaxation processes. One is the Markovian approximation (sections 2.2 and 2.4) and another is the rotating wave approximation (RWA) (section 5.4). While there are regions when these two approximations overlap, in the general case these approximations are independent ones. In the Markovian approximation we get conventional master equations (2.7.10), which for the case of the spontaneous emission of the two-level systems transform into a simple equation (a zero absolute temperature, T = 0)

$$\dot{P}_+ = -wP_+ ; \quad \dot{P}_+ = e^{-wt} ; \quad P_+(0) = 1 \ , \tag{5.7.1}$$

and w is the transition probability between states |+> and |->, while $P_+(t)$ is the probability to remain in the upper state |+>. We obtain the same result, (5.4.26), in the RWA. However, in the RWA there is the dissipationless region, (5.4.27), (5.5.9), in which the time-dependence is reversible (5.4.39) and the asymptotic value of $P_{\pm}^{\infty} \neq 0$. This is apparently a non-Markovian behavior. Thus, the RWA not necessarily implies the Markovian approximation.

The weak coupling approximation has main features of RWA. However, for the two-level system, the interaction energy in the RWA, (5.4.6), does not affect this lower level $E_g = 0$. Therefore the difference between exact upper and lower level energies equals $-\hbar\Delta_0$. In the weak coupling approximation, when counterrotating terms are taken into account, $E_g = -\hbar\Delta_c$, (5.5.28), and the difference between exact upper and lower eigenenergies has the opposite sign $E_d - E_g = \hbar\Delta_0$, (5.5.29).

It is expedient to compare various approximations (RWA, Markovian, weak coupling) using some exact solution. Such an exact solution exists for the harmonic oscillator interacting with the harmonic thermal bath – harmonic phonon (boson) dissipative system. A comparison between the rotating wave approximation (RWA) and the exact solution for the harmonic oscillator has been performed by Ondrechen, Nitzan, and Ratner [156], Lindenberg and West [157]. The undamped solutions have been taken into account by Cukier and Mazur [158].

We write the Hamiltonian of the system in the form ($\hbar = 1$)

$$H = \omega_0 a^+ a + \sum_{\nu} \omega_{\nu} b_{\nu}^+ b_{\nu} + \sum_{\nu} (G_{\nu} b_{\nu} + G_{\nu}^* b_{\nu}^+)(a + a^+) \ , \tag{5.7.2}$$

where the first two terms represent the uncoupled harmonic oscillator and harmonic thermal bath, and the last term represents their interaction. Solving differential equations for each of the operators a and b_{ν} by performing the Laplace transform, one obtains

$$a(t)=\frac{1}{2\pi i}\int_{s-i\infty}^{s+i\infty} e^{st}a(s)ds=\frac{1}{2\pi i}\int_{s-i\infty}^{s+i\infty} e^{st}\Big[U_1(s)a(0)+U_2(s)a^+(0)$$

$$+\sum_v\left(V_{1v}(s)b_v(0)+V_{2v}(s)b_v^+(0)\right)\Big]ds \quad , \tag{5.7.3}$$

where

$$U_1(s)=\frac{s-i\omega_0+R(s)}{s^2+\omega_0^2+2i\omega_0R(s)} \quad , \tag{5.7.4}$$

$$U_2(s)=\frac{R(s)}{s^2+\omega_0^2+i\omega_0R(s)} \quad , \tag{5.7.5}$$

$$V_{1v}(s)=\frac{s-i\omega_0}{s^2+\omega_0^2+2i\omega_0R(s)}\,\frac{iG_v}{s+i\omega_v} \quad , \tag{5.7.6}$$

$$V_{2v}(s)=\frac{s-i\omega_0}{s^2+\omega_0^2+2i\omega_0R(s)}\,\frac{iG_v^*}{s-i\omega_v} \quad , \tag{5.7.7}$$

$$R(s)=-\sum_v|G_v|^2\left[\frac{1}{s+i\omega_v}-\frac{1}{s-i\omega_v}\right] \quad . \tag{5.7.8}$$

Let us consider an average value of $<a(t)>$, assuming that initially the thermal bath was in equilibrium:

$$<b_v(0)>=<b_v^+(0)>=0 \quad . \tag{5.7.9}$$

For the average oscillator amplitude one gets an expression

$$<a(t)>=I(t)+\sum_k e^{z_kt}\,\mathrm{Res}(z_k) \quad , \tag{5.7.10}$$

where $z_k = \pm i\omega_k$ represent nondecaying local modes or (as we will see below) $z_k = \pm x_0$ represent nonoscillating modes, and I(t) denotes the contour integral

$$I(t)=\frac{1}{\pi}\int_{\omega_1}^{\omega_2}\varphi(y)<a(0)>\frac{e^{iyt}(\omega_0-y)^2-e^{-iyt}(\omega_0+y)^2}{(\omega_0^2-y^2-2\omega_0F(y))^2+4\omega_0^2\varphi^2(y)}dy$$

$$+\frac{1}{\pi}\int_{\omega_1}^{\omega_2}\varphi(y)<a^+(0)>\frac{(e^{iyt}-e^{-iyt})(\omega_0^2-y^2)}{(\omega_0^2-y^2-2\omega_0F(y))^2+4\omega_0^2\varphi^2(y)}dy\,. \qquad (5.7.11)$$

Here

$$F(y)=-\fint_{\omega_1}^{\omega_2}|G(\omega)|^2\left[\frac{1}{y-\omega}-\frac{1}{y+\omega}\right]d\omega\ , \qquad (5.7.12)$$

$$\varphi(y)=\pi|G(y)|^2\chi_y(\omega_1,\omega_2)-\pi|G(-y)|^2\chi_y(-\omega_1-\omega_2)\ , \qquad (5.7.13)$$

$$|G(\omega_\nu)|^2=|G_\nu|^2\rho(\omega_\nu)\ , \qquad (5.7.14)$$

where $\rho(\omega)$ is the frequency density of the phonon bath. The $\fint$ sign in (5.7.12) denotes the principal value, ω_1 and ω_2 are minimum and maximum frequencies of the phonon bath, and $\chi_y(a,b)$ is the function, equals to 1 when $y\in(a,b)$ and equals zero otherwise. The zeros of the denominators in Eq. (5.7.11) can easily be found

$$y_0=\pm\sqrt{\omega_0^2-2\omega_0F\pm 2i\omega_0\varphi}\ . \qquad (5.7.15)$$

In the weak coupling approximation $|F|<<\omega_0$, $|\varphi|<<\omega_0$

$$y_0\approx\pm(\omega_0-F)\pm i\varphi\ . \qquad (5.7.16)$$

Assuming also that one can neglect dependence of $F(y)$ and $\varphi(y)$ on y in the vicinity of ω_0 (Markovian approximation), the integration can be expanded on the whole y-axes. In this case one can use the residue theorem and obtain the time dependence of mean values $<a(t)>$ and $<a^+(t)>$ in the form

$$e^{\pm i(\omega_0-F)t-\varphi t}\ . \qquad (5.7.17)$$

We now consider the second term of Eq. (5.7.10). It turns out [159] that if poles z_k exist, they are either purely real or pure imaginary. Let us consider these two cases separately.

If the poles have no real parts, one gets the modes without damping, i.e. local modes or local vibrations. These local modes may lie only outside the spectrum of the phonons, $|y|<\omega_1$ or $|y|>\omega_2$ (note that ω_1 may be zero and ω_2 infinity; in that case isolated undamped modes cannot exist).

Isolated modes exist if the following conditions are satisfied

$$\int_{\omega_1}^{\omega_2} |G(\omega)|^2 \frac{1}{\omega} d\omega < \omega_0/4 \ , \tag{5.7.18}$$

$$\int_{\omega_1}^{\omega_2} |G(\omega)|^2 \left[\frac{1}{\omega-\omega_1} - \frac{1}{\omega+\omega_1}\right] d\omega > \left(\omega_0^2 - \omega_1^2\right)/2\omega_0 \ , \tag{5.7.19}$$

for the pole in the region $|y| < \omega_1$

$$\int_{\omega_1}^{\omega_2} |G(\omega)|^2 \left[\frac{1}{\omega_2-\omega} - \frac{1}{\omega_2+\omega}\right] d\omega > \left(\omega_2^2 - \omega_0^2\right)/2\omega_0 \ , \tag{5.7.20}$$

for the pole in the region $|y| > \omega_2$.

In these cases the oscillating local modes have no damping. Hence, $\langle a(t) \rangle$ has no limit as $t \to \infty$; this reversible behavior could not be obtained in the Markovian approximation.

The contribution to the sum of residues (5.7.10) can also be made by the poles with zero frequency $y = 0$. In that case local vibrations do not exist and there are two solutions equal in absolute value and different in sign, $z = \pm x_0$,

$$\langle a(t) \rangle = I(t) + e^{x_0 t}\,\mathrm{Res}(x_0) + e^{-x_0 t}\,\mathrm{Res}(-x_0) \ . \tag{5.7.21}$$

The condition of existence of these poles can be written as

$$\int_{\omega_1}^{\omega_2} |G(\omega)|^2 \frac{1}{\omega} d\omega > \omega_0/4 \ . \tag{5.7.22}$$

Comparing this condition with (5.7.18) we see that the nonoscillating terms due to the poles (5.7.21) appear immediately after the disappearance of local oscillating modes.

Now the whole dynamics of the poles can be described. When coupling is small enough, none of the conditions (5.7.19), (5.7.20) and (5.7.22) is satisfied. As the coupling increases, two poles appear at the points $\pm i\omega_1$ and they approach the origin. (We do not consider poles connected with the upper level ω_2). The poles reach the origin when the inequality (5.7.18) turns into equality. After condition (5.7.22) is satisfied, the poles begin to move from the origin along the real axes. For such strength of coupling we obtain two nonoscillating solutions; one of them is decreasing and the other is increasing.

The existence of the nonoscillating modes and especially of the increasing mode, looks rather unusual. However, the instability of this kind can be obtained in much simpler cases, even in the problem of two interacting harmonic oscillators. This problem is, of course, classical and discussed in every book on mechanics (see, for example, Ref. 160). Nevertheless, the possibility of the existence of such solutions is not described there. If we consider two identical coupled oscillators with the Hamiltonian

$$H=\frac{1}{2}\left(p_x^2+p_y^2\right)+\frac{\omega_0^2}{2}\left(x^2+y^2\right)+\alpha xy \quad , \tag{5.7.23}$$

we will immediately obtain the eigenfrequencies of this system

$$\omega_1^2=\omega_0^2-\alpha\,; \qquad\qquad \omega_2^2=\omega_0^2+\alpha \quad . \tag{5.7.24}$$

Obviously, if the coupling constant is sufficiently large $\alpha>\omega_0^2$, ω_1^2 will be negative, that is we obtain increasing nonoscillating solutions. It should be noted that this solution does not contradict the law of energy conservation. The situation when one of the frequencies becomes imaginary, corresponds to the case when the potential energy is no longer described by a positive definite form. The exponential increase of the kinetic energy is compensated by the increase of the absolute value potential energy which has a negative sign.

When we consider a large number of oscillators, the situation is similar, but with one difference: in this configuration each oscillator of the thermal bath interacts with only one singled out oscillator with frequency ω_0. This means that the frequency shift of each of the oscillators ω_ν in the continuum is too small to cause the instability (its magnitude is proportional to N^{-1}, where N is the number of oscillators in the bath). The singled out oscillator (with frequency ω_0), on the contrary, is coupled to the whole continuum and due to this its frequency may become imaginary. Physically, it means that only anharmonic terms, if they exist, make the motion of the oscillator finite.

The above analysis may be applied to the harmonic oscillator chain with an impurity [158]. In this case the frequency of the singled out oscillator (ω_0 in our case) depends on the coupling constant and condition (5.7.22) cannot be satisfied.

Now we can summarize the conditions under which the exact solution (5.7.10 – 5.7.14) can be approximated by the Markovian solution.

(a) The relaxation term I(t) should have exponential behavior $e^{-\gamma t}$. It takes place in the pole approximation, when the denominator in Eq. (5.7.11) may be approximately presented as

$$\Im=(\omega_0^2-y^2-2\omega_0F(y))^2+4\omega_0^2\varphi^2(y)\approx\Im'(y_0)(y-y_0) \quad , \tag{5.7.25}$$

with y_0 given by Eq. (5.7.15). This approximate equality takes place, provided

$$\varphi<<\frac{\Im'(y_0)}{\Im''(y_0)}\equiv\omega^*=\tau_c^{-1} \quad , \tag{5.7.26}$$

where ω^* and τ_c are quantities appearing in sections 1.7 , Eq. (1.7.16), and sections 2.2 – 2.4. Exact expressions for y_0 is given by Eq. (5.7.15), while the weak coupling expression is given by Eq. (5.7.16). It is worthwhile to mention that the weak coupling is not a prerequisite of the Markovian approximation; a necessary condition of the latter is given by Eq. (5.7.26).

(b) In the Markovian approximation local vibrations and nonoscillating terms (5.7.21) should be absent. Therefore the necessary condition of the Markovian approximation is the nonfulfillment of the inequalities (5.7.19) and (5.7.20). It can be shown that local vibrations with a frequency higher than ω_2 usually appear when local vibrations with a frequency lower than ω_1 already exist. Hence, the condition of the absence of isolated modes (and nonoscillating ones) take place

$$\int_{\omega_1}^{\omega_2} |G(\omega)|^2 \frac{1}{\omega} d\omega < \omega_0/4 \ , \tag{5.7.27}$$

$$\int_{\omega_1}^{\omega_2} |G(\omega)|^2 \frac{\omega d\omega}{\omega^2-\omega_1^2} < \left(\omega_0^2-\omega_1^2\right)/4\omega_0 \ . \tag{5.7.28}$$

If these inequalities are satisfied but the two sides of it are of the same order, the relaxation is still nonexponential. Hence this condition of Markovian approximation has to take the form

$$\int_{\omega_1}^{\omega_2} |G(\omega)|^2 \frac{1}{\omega} d\omega << \omega_0/4 \ , \tag{5.7.29}$$

$$\int_{\omega_1}^{\omega_2} |G(\omega)|^2 \frac{\omega d\omega}{\omega^2-\omega_1^2} << \left(\omega_0^2-\omega_1^2\right)/4\omega_0 \ . \tag{5.7.30}$$

These inequalities are independent of (5.7.26).

It has to be stressed that all the above results are temperature-independent. This means that these results are valid both for the high temperatures and for the vacuum state. The temperature dependence appears when we consider the relaxation of higher powers of the operator a: aa^+, a^2, and so on.

Now we consider the rotating wave approximation. Within the RWA the Hamiltonian (5.7.2) is reduced to

$$H = \omega_0 a^+ a + \sum_\nu \omega_\nu b_\nu^+ b_\nu + \sum_\nu \left(G_\nu b_\nu a^+ + G_\nu^* b_\nu^+ a\right) \ . \tag{5.7.31}$$

The RWA Hamiltonian (5.7.31) generates two independent sets of equations for a^+ and a, while the full Hamiltonian generates the interdependent equations for a^+ and a. Thus equations of motion for a^+ and b_j^+ have the form

$$\dot{a}^+ = i\omega_0 a^+ - i\sum_\nu G_\nu^* b_\nu^+ \ , \tag{5.7.32}$$

$$\dot{b}_\nu^+ = i\omega_\nu b_\nu^+ + iG_\nu a^+ \ . \tag{5.7.33}$$

Similar equations are obtained for a and b_v by performing the Hermitian conjugation of Eqs. (5.7.32 -5.7.33). Equations for the averaged operators $<a^+>$ and $<b_v^+>$ are linear and have the same form as Eqs. (5.7.32) and (5.7.33). Like in the general case, the equations for the averaged $<a^+>$ and $<b_v^+>$ do not depend on the bath temperature, and they are valid for the vacuum state. This means that if initially the oscillator was in the ground state n = 0, it will remain in this state forever. This is the zero solution $<a^+> = 0$ of the equations for averaged $<a^+>$. If the wave function of oscillator a^+ is a superposition of states n = 0,1 only, then the Hamiltonian (5.7.31) generates transitions $n=1 \rightleftarrows n=0$, while the transitions to the higher states n > 1 are forbidden. In this the RWA Hamiltonian differs from the full Hamiltonian (5.7.2), which connects all quantum numbers n together.

Considering the two-level system n = 0;1, one can identify components r_1, r_2 and r_3 with the following combinations of a and a^+

$$a^+a = r_3 + \frac{1}{2}; \quad a^+ = r_+; \quad \text{and} \quad a = r_- . \tag{5.7.34}$$

Now the Hamiltonian (5.7.31) coincides with the two-level RWA Hamiltonian (5.4.5), where $\Delta_0 = \omega_0$ and $\hbar = 1$, $a_k = b_v$, $B_k = G_v$ and $\omega_k = \omega_v$.

Representing operator $a^+(t)$ in the form

$$a^+(t) = a^+ e^{i\omega t} , \tag{5.7.35}$$

we obtain the equation for ω

$$\omega = \omega_0 + \sum_v \frac{|G_v|^2}{\omega - \omega_v} = \omega_0 + \int_{\omega_1}^{\omega_2} \frac{|G(y)|^2}{\omega - y} dy , \tag{5.7.36}$$

which is identical with Eq. (5.4.10). For the sake of simplicity we will consider the case $\omega_1 = 0$; $\omega_2 = \infty$. Like in section 5.4 we consider two cases. The first case is when ω_0 is larger than the critical frequency ω_c

$$\omega_0 > \omega_c = \int_0^\infty \frac{|G(\omega)|^2}{\omega} d\omega . \tag{5.7.37}$$

In this case there are no local modes and time dependence of $<a^+(t)>$ takes the form

$$<a'(t)> = \frac{1}{\pi}\int_0^\infty d\omega <a^+(0)> \frac{|G(\omega)|^2 e^{i\omega t}}{(\tilde{\omega}_0 - \omega)^2 + |G(\omega)|^2} , \tag{5.7.38}$$

where

$$\tilde{\omega}_0(\omega) = \omega_0 - \fint_0^\infty \frac{|G(y)|^2}{y-\omega} dy \equiv \omega_0 - F_{RWA}(\omega) \quad , \tag{5.7.39}$$

and $\fint$ means the principal value of the integer.

In the Markovian approximation, when one can neglect the ω-dependence of $\tilde{\omega}_0(\omega)$ and $|G(\omega)|^2$ in the vicinity of $\tilde{\omega}_0(\omega_0)$, we obtain the weak coupling case time dependence (5.7.17)

$$< a^+(t) > = < a^+(0) > e^{i(\omega_0 - F(\omega_0))t - |G(\omega_0)|^2 t} \quad . \tag{5.7.40}$$

In the opposite to (5.7.37) case

$$\omega_0 < \omega_c = \int_0^\infty \frac{|G(\omega)|^2}{\omega} d\omega \quad , \tag{5.7.41}$$

there is a local (non-decaying) mode with a negative frequency

$$\omega = -\Omega ; \qquad \Omega > 0 \quad , \tag{5.7.42}$$

which satisfies Eq. (5.7.36) (or similarly Eq. (5.4.29))

$$\Omega + \omega_0 = \int_0^\infty \frac{|G(\omega)|^2}{\Omega + \omega} d\omega \quad . \tag{5.7.43}$$

This mode is an artifact of the RWA. The exact Hamiltonian (5.7.2) generates nonoscillating exponentially increasing and decreasing modes, provided condition (5.7.22) is satisfied ($\omega_1 = 0$, $\omega_2 = \infty$)

$$\omega_0 < 4\omega_c \quad . \tag{5.7.44}$$

It means that in the region (see (5.7.37))

$$\omega_c < \omega_0 < 4\omega_c \quad , \tag{5.7.45}$$

the RWA Hamiltonian (5.7.31) leads to the decaying oscillating $<a^+(t)> \to 0$, when $t \to \infty$, (5.7.40), while in the region (5.7.41) the RWA gives the nondecaying local mode with negative frequency, satisfying Eq. (5.7.43). The exact solution in the region (5.7.45) does not contain nondecaying local modes. The exact solution in this region is (5.7.21); it contains nonoscillating decreasing and increasing modes. Therefore the RWA cannot be used in the region (5.7.44) and, as a matter of fact, in a broader region

$$\omega_0 \lesssim 4\omega_c \quad . \tag{5.7.46}$$

Only at the frequencies

$$\omega_0 >> \omega_c \ , \tag{5.7.47}$$

and in the weak coupling approximation, the exact Hamiltonian, (5.7.2), and the RWA Hamiltonian (5.7.31) give the same results.

Indeed the Markovian approximation and the RWA are independent ones. Both exact solutions (5.7.10) and (5.7.11) and the rotating wave solution (5.7.38) may be treated in the Markovian approximation (see conditions (a) and (b) on p. 166).

5.8 Impossibility of exponential relaxation

Both exact and RWA Hamiltonians lead to the exponential decay in the Markovian approximation (5.7.17), and (5.7.40). It should be stressed that these are approximate results. Khalfin [161] has shown that the exact exponential relaxation cannot be realized in physical systems. This result has been derived for pure states of the physical system. However, as is clear from the general considerations of Chapter II, the relaxation processes usually take place in mixed states, and are described by the density matrices.

The time dependence of the density matrix may be presented in a general case as (5.1.13)

$$\rho_{uv}(t) = \sum_{u'v'LM} S_{uL} S^{*}_{u'L} \rho_{u'v'}(0) S_{v'M} S^{*}_{vM} e^{-i\omega'_{LM} t} \ , \tag{5.8.1}$$

where S is a unitary matrix that connects the density matrix in the H_0-representation (see Hamiltonian (5.1.1)), with the density matrix in the H representation, $\hbar\omega'_{LM} = E'_L - E'_M$; E'_L, E'_M are the eigenvalues of the Hamiltonian H, and $\rho_{u'v'}(0)$ is the initial density matrix in the H_0 representation.

Let us transform the initial density matrix $\rho_{u'v'}(0)$ to the diagonal form [159]. It can be done by a unitary matrix T: $\tilde{\rho}(0) = T^{+}\rho(0)T$, where $\tilde{\rho}_{kl}(0) = \tilde{\rho}_{kk}(0)\delta_{kl}$. Then $\rho(0) = T\tilde{\rho}(0)T^{+}$, or $\rho_{u'v'}(0) = \sum_k T_{u'k}\tilde{\rho}_{kk}(0)T^{*}_{kv'}$. Substituting this to the diagonal part $\rho_{uu}(t)$, (5.8.1), we obtain

$$\rho_{uu}(t) = \sum_{u'v'L,M,k} S_{uL} S^{*}_{u'L} T_{u'k} \tilde{\rho}_{kk}(0) S_{v'M} S^{*}_{uM} T^{*}_{kv'} e^{-i\omega'_{LM} t} \ . \tag{5.8.2}$$

Separating the summation with respect to k, $(u'L)$, and $(v'M)$ we get

$$\rho_{uu} = \sum_k \tilde{\rho}_{kk}(0) |F_{uk}(t)|^2 \ , \tag{5.8.3}$$

where

$$F_{uk}(t) = \sum_{u'L} S_{uL} S^{*}_{u'L} T_{u'k} e^{-i\omega'_L t} , \tag{5.8.4}$$

and $\omega'_L = E'_L / \hbar$.

We will show now that $\rho_{uu}(t)$, the probability for the system to be found in the excited state u, tends to zero, as $t \to \infty$, is slower than the exponential function. Since all the values of $\tilde{\rho}_{kk}(0)$ are positive, we can always find an integer N, such that $0 < \lambda \equiv \tilde{\rho}_{NN}(0) \leq \tilde{\rho}_{kk}(0)$ for any k. Hence

$$\rho_{uu}(t) \geq \lambda \sum_{k} | F_{uk}(t) |^2 \ . \tag{5.8.5}$$

Denoting $\sum_{u'} S^{*}_{u'L} T_{u'k}$ by $\tilde{S}_{Lk}$ we obtain

$$F_{uk}(t) = \sum_{L} S_{uL} \tilde{S}_{Lk} e^{-i\omega'_L t} \ .$$

Passing from the sum to the Fourier integral we can finally write

$$F_{uk}(t) = \int_0^{\infty} w_{uk}(\omega) e^{-i\omega t} d\omega \ , \tag{5.8.6}$$

where $w_{uk} = S_u(\omega)\tilde{S}_k(\omega)f(\omega)$, and $f(\omega)$ is a density of frequencies ω'_L.

It follows from the theory of the Fourier series that the Fourier transform of the function with a bounded carrier cannot be an exponential function. Indeed, by the theorem of Paley and Wiener [163] and since the function $w_{uk}(\omega)$ can be defined as zero for negative ω, its Fourier transform $F_{uk}(t)$ must satisfy the inequality

$$\int_{-\infty}^{\infty} \frac{| \ln | F_{uk}(t) ||}{1+t^2} dt < \infty \ . \tag{5.8.7}$$

The implication of this condition is that $|F_{uk}(t)|$ cannot obey the exponential law of relaxation for all times. Indeed, if $|F_{uk}(t)|$ is an exponential function for $t \to \infty$, then integral (5.8.7) does not converge. This means that $|F_{uk}(t)|$ must vanish slower than the exponential function, at least as $\exp[-\gamma t/(\ln t)^{\alpha}]$, $\alpha > 0$. Coming back to (5.8.5), we see that $\rho_{uu}(t)$ possesses the same property. This means that pure exponential decay for a real physical system can never be realized.

It has to be emphasized that this statement relates to exact solutions, or exact soluble models. Obviously, an exponential relaxation can be obtained as an approximate solution of a problem.

CHAPTER VI

QUANTUM MEASUREMENT AND IRREVERSIBILITY

In previous chapters, various irreversibilities in quantum mechanics have been considered. Still, one can get an impression that the irreversible process is, in a sense, a foreign body in quantum mechanics. Typical, finite quantum systems, such as a two-state system, a harmonic oscillator, atoms, molecules are reversible. More than that, there are large quantum systems, like superconductors, superfluids and others, which are also reversible. In the previous chapter it has also shown that in even generally dissipative system with large ($N \to \infty$) number of degrees of freedoms, there is a certain range of parameters in which the system is reversible. Thus one may conclude that irreversible processes in quantum mechanics play an important, albeit not dominant, role. Though there are many reversible quantum-mechanical systems, such an impression does not reflect a real state of affairs.

Quantum measurement – the process and mechanism, lies in the very heart of quantum mechanics. The process of quantum measurement, even of simple systems such as the two-level molecule, has an irreversible character. Reduction of the wave packet, a collapse of the wave function is an irreversible process. Therefore, the irreversibility penetrates to the heart of quantum mechanics through the central role of quantum measurement in quantum mechanics.

In this chapter the process of quantum measurement is analyzed as a special case of the interaction of quantum system. The collapse of the wave function is considered as a result of the irreversible interaction between the coherent dynamic system described by the off-diagonal density matrix and the non-coherent dissipative system described by the almost diagonal density matrix.

6.1 Another view on quantum mechanics. EPR. Bell's theorems. Nonlocality.

In the first chapter we have presented a conventional (Copenhagen) description of quantum mechanics. A certain deviation from the conventional picture is that basic quantity describing the state is the density matrix, while the wave function is a specific case. In the conventional representation a strangeness and even mystery of quantum mechanics is, somehow, overlooked. In this section we wish to present another, nonpragmatic, view on quantum mechanics. The widely spread opinion, especially among philosophers, is that quantum measurements, the act observation, is something almost mystical – the object does not exist without being observed. This opinion stems from the influence of the very authoritative, and dominant, personality of Niels Bohr. The conventional Copenhagen interpretation is, as a matter of fact, Bohr's interpretation. Many physicists, taking their lead from the central figure of Niels Bohr, deny the reality of the quantum object before the measurement (or between the measurements). These physicists would say that there is no objective picture at all. Nothing is actually "out there", at the quantum level. Quantum theory, according to this view, is merely a calculation procedure that does not attempt to describe the world as it actually is. Thus Bohr states: "I warned especially against phases often found in physicists, such as 'creating physical attributes to atomic objects by measurements'." Such phrases are "apt to cause confusion..." [163]. Bohr is evidently saying here essentially what we have said above, i.e., that it has no meaning to talk of a quantum object with its attributes apart from the unanalyzable whole phenomenon in which it is actually observed.

Bohr's view had a very widespread influence, but his ideas do not appear to have been well understood by the majority of physicists. Rather, the latter generally followed a quite different approach initiated by Dirac and von Neumann, in which a quantum state played a central role. In the first chapter we used this concept without making a special emphasis on it. The wave function (or more generally the density matrix) gives the most complete possible description of quantum reality, which is thus contained in the concept of a quantum state. The evolution of a quantum object is described by the wave time-dependent function (or density matrix), while the process of the measurement (as will be shown below) can be described as a special case of quantum mechanical interaction.

Before considering the nature of the measurement process we will try to bring to the reader the unusualness of quantum theory. A deeper insight into the structure of quantum mechanics reveals a magnificent and counterintuitive picture, which contradicts any experience obtained by the physicist in other theories. In this context I cannot refrain from quoting Albert Einstein. "His [scientist's] religious feeling takes the form of a rapturous amazement at the harmony of natural law, which reveals an intelligence of such superiority that compared with it, all the systematic thinking and actions of human beings is an utterly insignificant reflection" [164]. Three thousand years ago the author of psalms expressed a similar feeling. "O Lord, how great are thy works! And thy thoughts are very deep. A brutish man does not know. Nor does a fool understand this" (Psalms, 92:67).

The EPR paradox

In 1935 Einstein, Podolsky and Rosen [165] wrote their famous paper, known as EPR. They do not question correctness of quantum mechanics. EPR ask the question whether quantum-mechanical description of physical reality is complete. To answer this question the notion of completeness has to be defined. According to EPR two conditions are necessary:

(1) *Every element of the physical reality must have a counterpart in the physical theory.*

(2) *If, without in any way disturbing a system, one can predict with certainty (i.e., with the probability $p = 1$) the value of a physical quantity, then there exists an element of physical reality corresponding to this physical quantity.*

Let us consider correlated quantum systems of type (1.4.7), which interact during a finite time interval. After the interaction between two systems 1 and 2 is vanished, we have

$$\psi(x_1, x_2) = \sum_{nk} c_{nk} u_n(x_1)\psi_k(x_2) \equiv \sum_n \varphi_n(x_2) u_n(x_1);$$
$$\varphi_n(x_2) \equiv \sum_k c_{nk}\psi_k(x_2) \quad . \tag{6.1.1}$$

Using properties of these correlated quantum systems, EPR concluded that quantum theory does not provide a complete description of physical reality. Let us follow their argumentation in order to verify whether the two above conditions and expansion (6.1.1) are enough to prove incompleteness of the quantum theory.

Let a_1, a_2, a_3,..., be the eigenvalues, and $u_1(x_1)$, $u_2(x_1)$,... be eigenvalues and eigenfunctions, respectively, of some physical quantity A pertaining to system A. Variables x_2 are used to describe the second system. Functions $\varphi_n(x_2)$ may be regarded as the coefficients of the expansion of ψ into a series of orthogonal functions $u_k(x_1)$. Suppose that the quantity A is measured and it is found that it has the value a_k. Then the *reduction* of the wave packet takes place. (In the next section a more detailed description of this reduction is given.) The infinite series (6.1.1) is reduced to a single term $\varphi_k(x_2)u_k(x_1)$. It is then concluded that after the measurement the first system is in the state given by the wave function $u_k(x_1)$, while the second system is in the state $\varphi_k(x_2)$.

The set of functions $u_n(x_1)$ is determined by the choice of the physical quantity A. Choosing another quantity, say B, with eigenvalues b_1, b_2,..., and eigenfunctions $v_1(x_1)$, $v_2(x_1)$,..., we get another expansion (representation) of the same wave function

$$\psi(x_1, x_2) = \sum_s \chi_s(x_2) v_s(x_1) \quad . \tag{6.1.2}$$

Measuring quantity B we find the whole system in the state $\chi_\ell(x_2)v_\ell(x_1)$. Thus the first system is in the state $v_\ell(x_1)$, while the second system is in the state $\chi_\ell(x_2)$. "We therefore see that, as a consequence of two different measurements performed upon the first system, the second system may be left in states with two different wave functions $\varphi_k(x_2)$ and $\chi_\ell(x_2)$. On the other hand, since at the time of measurement the two systems no longer interact, no real change can take place in the second system in consequence of anything that may be done to the first system. This is, of course, merely a statement of what is meant by the absence of interaction between the two system." [165] (Emphasis is mine, B.F.). Thus it is possible to assign two different wave functions $\varphi_k(x_2)$ and $\chi_\ell(x_2)$ to the same reality. These two wave functions φ_k and χ_ℓ may be eigenfunctions of two noncommuting operators corresponding to some physical quantities P and Q, respectively. Such a situation cannot be described by quantum theory. This means that quantum theory is incomplete.

The arbitrary point in the above derivation is the assumption that the measurement performed on the first system does not disturb the second system. In another place [166] Einstein puts this assumption in very clear-cut form: "But on one supposition we should, in my opinion, absolutely hold fast: the real factual situation of the system S_2 is independent of what is done with the system S_1, which is spatially separated from the former". This statement looks as a diktat to Nature to use *a-priori* philosophical considerations. On the other hand, it does not matter how obvious may be certain statements, they have to be supported by the experiment and measurement.

EPR formulate this in the following words: "The elements of the physical reality cannot be determined by *a-priori* philosophical considerations but must be found by an appeal to results of experiments and measurements." [165] As a matter of fact, EPR make the *a-priori* assumption that an action on distance cannot take place. Although our whole intuition says that physical theory should be <u>local</u>, it is, still, an arbitrary assumption. More than that, and it remained unnoticed for many years, that quantum mechanics is not a local theory. A long time ago, before the development of quantum theory connected with Bell's theorems, Einstein realized that quantum theory is intrinsically nonlocal [167]. It requires some reflection to see the nonlocality of quantum mechanics. In the example (6.1.1) the distance between 1 and 2 is long enough, so that one can neglect the interaction energy between them. However, the measurement performed on the first system changes the state of system 2:

$$\sum_n \varphi_n(x_2)u_n(x_1) \to \varphi_\ell(x_2)u_\ell(x_1) \quad . \tag{6.1.3}$$

This is a routine situation in quantum theory. Now, when EPR state that there is no interaction between the two systems 1 and 2, they mean that potential energy $V(r_1,r_2)$ of the interaction between these systems tends to zero. However, in quantum theory there is a long distance, nonlocal quantum interaction of the (6.1.3) type, i.e. the reduction of wave packet (6.1.1) as a result of the measurement performed on system 1, is an intrinsic feature of quantum theory.

These are general considerations. EPR provide a specific example. But Bohm's Gedanken experiment [168] has become much more popular. Consider a pair of spin one-half particles formed somehow in the singlet spin state and moving freely in opposite directions (see Fig. 8). The singlet wave function of these two particles has the form (1.6.23)

$$|\psi\rangle = \frac{1}{\sqrt{2}}\left(|\alpha,+\vec{n}\rangle|\beta,-\vec{n}\rangle - |\alpha,-\vec{n}\rangle|\beta,+\vec{n}\rangle\right) \quad , \tag{6.1.4}$$

where $|\alpha,+\vec{n}\rangle$ is a single particle state of the particle α having spin up (+) or down (−) along the direction of unit vector $\vec{n}$, which can point in any direction. Following the EPR logic, the spin of the particle β in any direction can be predicted with certainty from a measurement of the spin of particle α in that direction. If α has spin up (down) in direction $\vec{n}$, then β has spin down (up). According to EPR the spin of particle β is an element of physical reality, because we can predict with certainty its value without making a measurement on it ("without disturbing it"). Measuring spin α, first in the x direction, and then in the y direction (it may be another pair of spins, but in the same

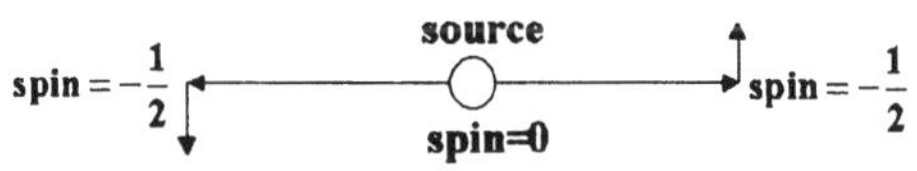

Fig. 8. A pair of spin ½ particles moving in opposite directions.

state (6.1.4)), one concludes that the spin component of particle β in both x and y directions must be elements of physical reality. According to the EPR suppositions, particles must, therefore, be moving along with definite values of its x and y spin components before any measurement is made. Now comes the EPR paradox: Quantum mechanics cannot assign simultaneous values to x and y spin components because the corresponding operators do not commute

$$[s_x, s_y] \neq 0 \ . \tag{6.1.5}$$

Therefore, according to EPR, quantum theory does not account for these elements of physical reality and therefore it cannot be considered a complete description of physical reality.

It is worthwhile to emphasize that there is no paradox within the framework of quantum theory itself, since EPR argument does not involve simultaneous measurements, or predictions of both x and y component particles b in the Bohm-Gedanken experiment. Rather, it involves the possibility of measuring either but not both, and since it is then possible to predict either with certainty, it is concluded that both are elements of physical reality.

Bell's Inequalities

Thus the locality is the main issue in the EPR paradox. The assumption of locality leads to the paradox and proof of the incompleteness of quantum mechanics. Bell [169] gave a very convincing answer to this problem. Figure 9 provides a schematic representation of the modification of Bohm's Gedanken experiment [168].

We choose units such that the observed spin in any given direction is either 1 or -1 ($\frac{1}{2}\hbar = 1$). At stations A and B there are instruments which measure the spin along an axis transverse to the line of flight of the particles. In each detector this transverse direction can be set in either of orientations θ at station A and ψ at station B. It is an empirical fact that, whenever the spin of the particle is measured along a given axis, one always finds that the spin points either "up" (+1) or "down" (−1) along that axis, but never to have some fractional value for its projection along an axis. That is, each individual measurement yields either +1 or −1, never any other value. The difference of the arrangement on Fig. 6.2 from the original Bohm's version of EPR is that two spins are not necessarily parallel, they may be measured with ψ-θ difference between their orientations.

Bell [169] formulated the criterion of the validity of local theory (not necessarily quantum) describing the experiment on Figure 9. He did this by considering the generalized version of the EPR experiment, representing the states of reality by a set of additional (hidden) variables λ. In the general case each individual result of the measurement of the projection of the spin on the axis $\vec{a}$ at station A, and on the axis $\vec{b}$ at station B, depend on the apparatus hidden variables μ_a or μ_b, variables λ_A and λ_B belonging respectively to particles A and B themselves, and a set λ which may be associated with the observed system as a whole. The results of the measurements are designated A = +1 for positive spin, and A = -1 for negative spin, and corresponding designations for spin B. Then in the general case

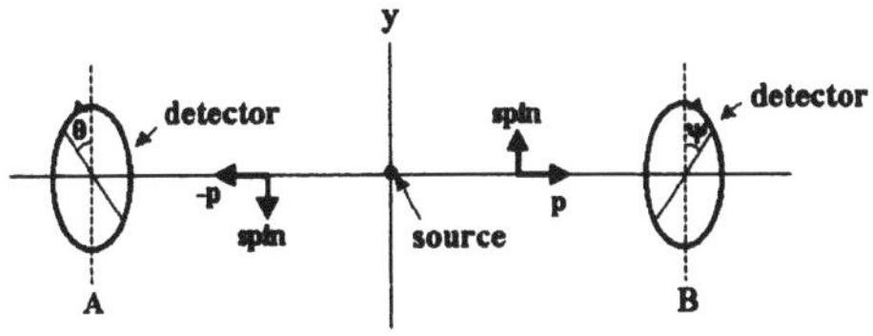

Fig. 9. At the center there is a source which decays and emits two electrons (or photons) in opposite directions with moments ±p. The spin of the atom ("source") is zero. At stations A and B there are instruments ("detectors" in the diagram) which can be set to measure the spin of the electron along an axis transverse to the line of flight of the electrons, at angles θ and ψ at stations A and B respectively.

$$A = A\left(\vec{a}, \mu_a, \vec{b}, \mu_b, \lambda_A, \lambda_B, \lambda\right) \ ,$$

$$B = B\left(\vec{a}, \mu_a, \vec{b}, \mu_b, \lambda_A, \lambda_B, \lambda\right) \ . \tag{6.1.6}$$

The interaction between A and B will be local if the result A depends only on μ_a and λ_A and not on μ_b and λ_B (and vice versa for B). Therefore

$$A = A(\vec{a}, \mu_a, \lambda_A, \lambda) \quad \text{and} \quad B = B(\vec{b}, \mu_b, \lambda_B, \lambda) \ . \tag{6.1.7}$$

Designating all hidden parameters by λ we may write

$$A(\vec{a}, \lambda) = \pm 1 \ ; \qquad B(\vec{b}, \lambda) = \pm 1 \ . \tag{6.1.8}$$

The vital assumption is that the result B for particle B does not depend on setting $\vec{a}$, of the magnet determining orientation of particle A and the same for particle B.

If $\rho(\lambda)$ is the probability distribution of λ, then the expectation value of the product of two components of spins A and B is

$$P(\vec{a}, \vec{b}) = \int d\lambda \rho(\lambda) A(\vec{a}, \lambda) B(\vec{b}, \lambda) \ . \tag{6.1.9}$$

Using (6.1.8) and (6.1.9) the following condition can be obtained

$$| P(\vec{a}, \vec{b}) - P(\vec{a}, \vec{c}) | + | P(\vec{d}, \vec{c}) + P(\vec{d}, \vec{b}) | \leq 2 \ . \tag{6.1.10}$$

This condition is Bell's inequality which must be satisfied for a local hidden theory to apply to our system of two particles with spin. (This symmetric form of the Bell inequality was first proposed by Clausner et al. [170].) These inequalities make possible a test for locality on the basis of measurements of four sets of correlations.

The prediction of quantum mechanics is

$$P(\vec{a},\vec{b}) = -\vec{a}\cdot\vec{b} \quad . \tag{6.1.11}$$

There is a range of angles for which (6.1.11) does not satisfy the inequality (6.1.10). Bell's inequality has been tested in a large number of experiments and generally speaking the inequality has been found to be violated. The most thorough set of experiments has been performed by Aspect et al. [171].

Two points have to be stressed. First, Bell did not write a local, deterministic theory. Rather, he proved that no such a theory can in principle exist. Second, Bell's theorem does not depend on quantum mechanics. It refutes a whole category of theories without ever mentioning quantum mechanics. It turns out that the experimental results not only refute the class of local deterministic theories, but also support nonlocal predictions of quantum mechanics.

Delayed-Choice Experiment

The nonlocal nature of quantum mechanics is manifested in the delayed-choice-experiment [172], Figure 10.

Photons are sent from a certain source. If the two possible routes are exactly equal in length, then it turns out that there is a 100 per cent probability that the photon reaches the detector A lying in the direction of the photon's initial motion and a zero per cent probability that it reaches the other detector B – the photon is certain to strike the detector A. It seems certain that the photon must, in some sense, have actually traveled both routes at once. For if an absorbing screen is placed in either one of the two routes, then it becomes equally probable that A or B is reached; but when both routes are open, only A can be reached. In a sense this experiment is very similar to the two-slit experiment. Closing one slit leads to the destruction of the interference pattern.

There are two differences. One is that in the delayed-choice experiment the interference is displayed when only one photon (or electron) is sent. In the two-slit experiment the interference pattern is displayed only when many electrons are sent. Thus the delay choice experiment manifests unequivocally that the interference is a property of one particle (electron, photon). Another difference is clearly manifested: nonlocality of the delayed choice experiment. Placing an absorbing screen on one of the routes, close to point L, at a time when photon is very close to its final destination N, leads to the destruction of the interference.

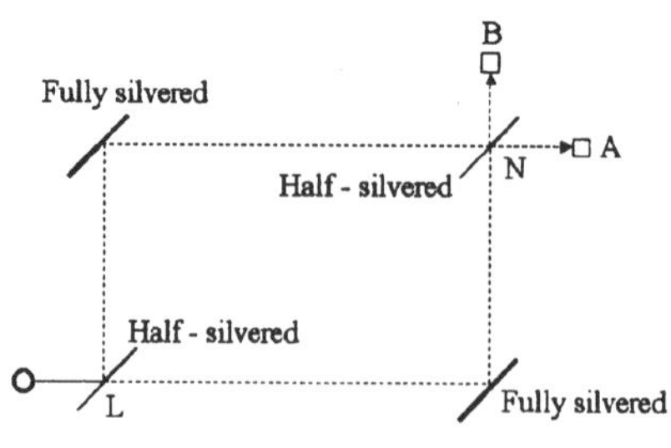

Fig. 10. The two routes taken by photon can be made to interfere with one another.

And, lastly, talking about the peculiar, nonlocal character of quantum mechanics, one can hardly avoid mentioning the de Broglie-Bohm causal interpretation of quantum mechanics. John Bell [169] writes:

> Bohm's 1952 papers [see Ref. 192] on quantum mechannics were for me a revelation...I have always felt...that people who have not grasped the ideas of those papers...and unfortunately they remain the majority...are handicapped in any discussion of the meaning of quantum mechanics.

The basic assumption is that the particles are real local entities, that have position and momentum, even though we cannot determine the values of them simultaneously. These particles are beables, in Bell's terminology, as opposed to the observables in conventional formulation of quantum mechanics. These particles, having well defined positions, can move along well defined trajectories.

To find the trajectories of quantum particles it is required that these trajectories depend on the wave function

$$\vec{r} = \vec{r}(\psi, t) \quad , \tag{6.1.12}$$

while the wave function is presented in the form

$$\psi = R\exp(iS/\hbar) \quad . \tag{6.1.13}$$

Considering the standard procedure leading to the semiclassical (WKB) approximation, we obtain the quantum Hamilton-Jacobi equation

$$\frac{\partial S}{\partial t} + \frac{(\nabla S)^2}{2m} + U + Q = 0; \qquad \frac{dR^2}{dt} + \nabla \bullet \left(\frac{R^2 \nabla S}{m} \right) = 0 \quad , \tag{6.1.14}$$

where $P = R^2$ is the probability density. Here U is a usual potential, while Q is the quantum potential

$$Q = -\frac{\hbar^2}{2m} \frac{\nabla^2 R}{R} \quad . \tag{6.1.15}$$

The second equation (6.1.14) expresses the conservation of probability. The momentum of the particle has the form

$$\vec{p} = \nabla S \quad . \tag{6.1.16}$$

The next step is almost evident, though it has been overlooked for many years. Q is added to the usual potential in the equation of motion

$$m\frac{d^2\vec{r}}{dt^2}=-\nabla(U)-\nabla(Q) \quad . \tag{6.1.17}$$

The two-slit experiment may serve as an example of an application of equations (6.1.14) – (6.1.17). It has been shown [192] that ensemble of electrons moving according to the equation of motion (6.1.17) results in the conventional interference pattern in the two-slit configuration.

Now we will demonstrate the appearance of quantum nonlocality in the Bohm's formalism. Let us consider the two-body system. The wave function $\psi(\vec{r}_1,\vec{r}_2,t)$ satisfies quantum Hamilton-Jacobi equations (with equal masses)

$$\frac{\partial S}{\partial t}+\frac{(\nabla_1 S)^2}{2m}+\frac{(\nabla_2 S)^2}{2m}+U+Q=0 \quad , \tag{6.1.18}$$

$$\frac{dR^2}{dt}+\nabla_1\cdot(R^2\nabla_1 S/m)+\nabla_2\cdot(R^2\nabla_1 S.m)=0 \quad , \tag{6.1.19}$$

where

$$Q=-\frac{\hbar^2}{2m}\frac{(\nabla_1^2+\nabla_2^2)R}{R} \quad , \tag{6.1.20}$$

and subscripts 1 and 2 refer to the first and second particles respectively. Now the momenta of the particles are

$$\vec{p}_1=\nabla_1 S(\vec{r}_1,\vec{r}_2 t)\ ; \quad \vec{p}_2=\nabla_2 S(\vec{r}_1,\vec{r}_2 t) \quad , \tag{6.1.21}$$

and R is a function of the coordinates $\vec{r}_1$ and $\vec{r}_2$ of the particles. Thus we see, in general, that the trajectory of particle 1 depends on the trajectory of particle 2 and vice versa. Indeed, this interdependence is not uncommon for the system of two interacting particles. However, the important feature of the quantum potential is that the particles can be separated by a considerable distance and yet interact very strongly, even though there may be no classical potential U between them. Here we also have the possibility of a non-local force. Indeed, this is just the kind of force that can offer an explanation of the EPR correlations. For the (6.1.1) type wave functions (and (6.1.4)) the quantum potential Q does not vanish when the distance between particles is very large (and tends to infinity). Thus, although the particles are not interacting through any classical force and are separated in space, they are interacting through the quantum potential. The measurement of the spin of one particle influences the state of the other particle, in agreement with quantum mechanics.

In equation (6.1.18), U and Q appear in essentially the same way. The difference is that Q describes the interaction that depends on the details of the whole wave function. While in classical mechanics the whole picture is determined by the local interactions between the particles, the quantum interaction is non-local and depends itself on the state of the whole system. Of course, it can be easily shown, that for the factorized wave function of the (1.4.2) type, the whole interaction becomes local. It was exactly the appearance of this non-locality that led John Bell to think about the problem of non-locality in more general terms and subsequently produced his inequalities.

Going on to the N-body system, we have

$$\psi = \psi(\vec{r}_1, \vec{r}_2, \ldots, \vec{r}_N, t) \quad , \tag{6.1.22}$$

and

$$Q = Q(\vec{r}_1, \vec{r}_2, \ldots, \vec{r}_N, t) \quad , \tag{6.1.23}$$

so that the behavior of each particle may depend nonlocally on the configuration of all others, no matter how far away they may be. As in the one-body case, we may take

$$P = R^2 \quad , \tag{6.1.24}$$

as the probability density, but this is now in the configuration space of all particles. Similar to Eq. (6.1.13) a classical particle moving in the electromagnetic field depends on this field. The difference is that the electromagnetic field is the function in the usual three-dimensional space, while ψ, (6.1.22), and Q, (6.1.23), are functions in the many-dimensional configuration space of all particles. In a sense, one may say that quantum mechanics is a local theory in many-dimensional configuration space and is a nonlocal one in the usual three-dimensional space.

All the above examples and considerations demonstrate a nonconventional and counterintuitive character of quantum mechanics. What is common to the considered processes is the decisive role of the measurement and reduction of the wave-packet.

6.2 Reduction of wave-packet

In a sense, the measurement process plays a central role in quantum mechanics (or, at least, in its interpretation). This problem continues to attract the attention and interest of researchers up till now. Classical works on the subject are given in the book [173]. Among them are papers of Wigner, EverettIII; Daneri, Loinger and Prosperi (to this should be added a paper of Rosenfeld [174]), Zeh and others. Other works including those of Zurek, Leggett and others are reviewed in Refs. [175, 176]).

An important development of the theory of quantum measurement is connected with the formalism of decoherent histories (DH), (Aharonov, Bergman and Lebowitz, Griffiths, Unruh, Gell-Man and Hartle, and Omnés). This development is reviewed in Refs. 173, 175 and 176. While DH analyze a sequence of measurements, the subject of the following sections is the description and understanding of a single act of measurement.

Suppose that a quantum system is described by the wave function

$$\psi = \sum c_n u_n \quad , \tag{6.2.1}$$

and the corresponding density matrix

$$\rho_{mn} = c_n^* c_m \quad . \tag{6.2.2}$$

In a more general case the system is described by the density matrix which does not coincide with (6.2.2), but also has nonzero off-diagonal matrix elements $\rho_{mn}(1-\delta_{mn})$. By measuring a dynamical quantity, say the energy, of which u_n are eigenfunctions, we obtain various eigenvalues $E_1, E_2,\ldots,E_n,\ldots$, with probabilities $P_n = |c_n|^2$, or $P_n = \rho_{nn}$ in a more general case. But once having obtained a given eigenvalue, say E_n (as a result of the measurement) we know that the system is necessarily in the state u_n. At the end of the measurement process the initial wave function ψ (the density matrix ρ) is transformed into a mixture of various pure states u_n with the probabilities $|c_1|^2$, $|c_2|^2,\ldots,|c_n|^2,\ldots$, or $\rho_{11},\ldots,\rho_{22},\ldots,\rho_{nn},\ldots$

The corresponding density matrix describing this proper mixture (see section 1.6) is

$$\rho_{mn} = \sum \rho_{nn} \,|\, u_n >< u_n \,|\, \delta_{mn}, \qquad \rho_{nn} = |\, c_n \,|^2 \quad , \tag{6.2.3}$$

or, in a general case, ρ_{nn} is a diagonal element of a general density matrix ρ.

The crucial difference between the density matrix (6.2.3), and wave-function (6.2.1) (or, more generally, the density matrix possessing off-diagonal matrix elements) is that coefficients of (6.2.3) may be interpreted as classical probabilities. The density matrix (6.2.3) can be used to describe the alternative states of the system. When off-diagonal terms are absent one can safely maintain that the system is in a definite, but unknown, state. Each particular state $|u_n>$ appears as a result of a certain (one) measurement, as a result of a transformation

$$\psi = \sum_n c_n u_n \rightarrow u_n \quad . \tag{6.2.4}$$

The multitude of a very large number of measurements is described by the density matrix (6.2.3), while ρ_{nn} are probabilities (relative frequencies) that transformation (6.2.4) takes place. When a measuring device performs the multitude of the measurements, the observer just verifies the relative frequencies of the outcome of the experiment. The density matrix of the (6.2.3) type can describe the result of throwing up the coin

$$\rho_{coin} = \frac{1}{2} |\, H >< H \,| + \frac{1}{2} |\, T >< T \,| \quad . \tag{6.2.5}$$

This correctly represents the certainty of two alternatives – that is whether the heads (H) or tails (T) are results of throwing the coins, while $\frac{1}{2}$ are relative frequencies of the experiment.

The above description assumes the passive role of the observer. He has only to verify the outcomes of the measurements. There is another point of view attributing to the observer an active role in the transformation (6.2.4). Below we will discuss this possibility in more detail.

The transformation (6.2.4) or transformation from (6.2.1) to (6.2.3) is often called the reduction of the wave-packet (or collapse of the wave function). Thus, at least at first sight, there are two entirely different processes. One of the processes is the evolution of the wave function (or density matrix) of the system between the measurements. This evolution is described by the unitary operator

$$U = e^{-\frac{i}{\hbar}Ht} , \qquad (6.2.6)$$

where H is the Hamiltonian of the system. On the other hand, there is another process – the quantum measurement. This process leads to the collapse of the wave function, or density matrix. Generally speaking the off-diagonal density matrix, whose time evolution is described by the unitary operator (6.2.6) is transformed by the measurement (collapse) to the diagonal density matrix. Such a transformation cannot be described by the unitary operator (6.2.6).

Von Neumann [177] was the first who distinguished two different processes: **1** – the measurement, and **2** – the unitary evolution of the state. Penrose [178] designated these two processes as **U** – the unitary evolution, and **R** – the measurement process. Regarding the density matrix, or wave function, as describing the "reality" of the system, we realize that the **U** process is entirely deterministic, while it is the procedure **R**, and only **R**, that introduces uncertainties and probabilities into quantum theory. The question arises whether the process **R** can be described by the usual quantum mechanics. Penrose [178] answers this question quite unequivocally:

> According to the standard procedures of quantum mechanism there is no implication that there be any way to 'deduce' **R** as a complicated instance of **U**. It is simply a *different* procedure from **U**, providing the other 'half' of the interpretation of the quantum formalism. All the non-determinism of the theory comes from **R** and not from **U**.

This conclusion seems inevitable once one assumes that a combined system containing an interacting measurable object and measuring device is described by the wave function. Suppose that we want to measure a quantity A of the object which is in the state

$$\psi = \sum c_k u_k(x) , \qquad (6.2.7)$$

where u_k is the eigenfunction corresponding to the value a_k of operator A, while x is the set of the variables of the object. Let M(y) be the coordinate specifying the position of the "pointer" of the measuring device, and m_0, m_1, m_2,...,m_p its eigenvalues, with eigenfunctions $v_0(y)$, $v_1(y)$,...,$v_p(y)$, where y is the set of the measuring device variables. Before the coupling we attribute to the combined system a collective wave function of the form

$$\psi(x,y) = v_0(y)\sum_k c_k u_k(x) \quad . \tag{6.2.8}$$

After the interaction the wave function obtains the form (cf. section 1.4)

$$\psi(x,y) = \sum_{k,\rho} c_{k\rho} u_k(x) v_\rho(y) \quad . \tag{6.2.9}$$

However, this form of the wave function implies that the pointer is in the superposition of its states. This is the situation of Schrödinger's "cat paradox" [179]. The introduction of another system (environment, observer), interacting with the combined object + measuring device, does not save the situation. The new wave function will take the form

$$\psi(x,y,z) = \sum_{k,\rho,\lambda} c_{k\rho\lambda} u_k(x) v_\rho(y) w_\lambda(z) \quad , \tag{6.2.10}$$

where $w_\lambda(z)$ are the eigenfunctions of the operator E of the third system (of the environment). At a certain stage there should be the collapse, (6.2.4), of the wave function

$$\psi = \sum_{k,\rho,\lambda} c_{k\rho\lambda} u_k(x) v_\rho(y), w_\lambda(z) \to u_k v_\rho w_\lambda \quad . \tag{6.2.11}$$

However, such a transformation contradicts the linear Schrödinger equation. On the other hand, it seems almost inevitable that such a transformation (the collapse of the wave function) must happen at a certain stage of the measurement. As von Neumann [177] pointed out, the point at which (6.2.11) occurs is not obvious, and he proposed that it would ultimately take place when a human consciousness (the "observer") is involved. A similar emphasis on consciousness has been relied by Wigner [180], Peierls [181], and London and Bauer [182].

The way of explaining a collapse of the wave function (6.2.4) or (6.2.11) is in postulating "that the equations of motion of quantum mechanics cease to be linear, in fact they are grossly non-linear if conscious beings enter the picture" [180]. Of course, all this approach removes the measurement problem into the region of the unknown. It implies that the destruction of the superposition is performed by the consciousness. I believe that the observer's mind has nothing to do with the measurement process. The measurement is a physical process of interaction of the system with the measuring device and environment. The role of the human being is to verify that the measurement took place. Various observers will see the same position of the pointer. They do not affect the measurement process itself. This point of view will be pursued in the next sections of this chapter.

6.3 Reduction of the wave packet as a result of interaction with the dissipative system

As we have mentioned in the previous section, the interaction with the object represents a measurement, provided that the reduction of the wave packet occurs, i.e., off-diagonal density matrix elements vanish. Von Neumann [177] postulated that, in addition to the unitary evolution (6.2.6), there is an ad hoc process **1** (or **R**) – a nonunitary reduction of the state vector – that transforms a pure, correlated state into an appropriate mixture. This may also take place when we consider an extended system, the object + measuring device, this extended system being initially in the pure system [175] (in a two-level system)

$$\rho = |\alpha|^2 |+><+||M_+><M_+| + \alpha\beta^* |+><-||M_+><M_-|$$

$$+\alpha^*\beta |-><+||M_-><M_+| + |\beta|^2 |-><-||M_-><M_-| \ . \qquad (6.3.1)$$

Here + and – are two states of the object and measuring device. Again the unitary evolution cannot reduce this density matrix to

$$\rho^r = |\alpha|^2 |+><+||M_+><M_+| + |\beta^2>|-><-||M_-><M_-| \ . \qquad (6.3.2)$$

It should be mentioned that the wave function reduction $\rho \to \rho^r$ (if it can be explained) does not provide the derivation of the probability interpretation. The statement that diagonal elements of the density matrix are the probabilities of certain states is one of the postulates of quantum mechanics. However, the consistency of the probability interpretation requires the reduction of the wave packet $\rho \to \rho^r$, since only in this reduced state the measuring device may have distinct states $|M_+>$ and $|M_->$, and not the superposition of them.

One may suppose that the interaction with the environment E, which is described by the wave function, may cause the reduction of the wave packet. Let us assume that initially the object + measuring device is in the state uncorrelated with the environment. Then the interaction process, described by the unitary transformation (6.2.6), leads to the correlated wave function

$$|\psi_{OD}>|E_0> = (\alpha|+>|M_+> + \beta|->|M_->)|E_0>$$

$$\to \alpha|+>|M_+>|E_+> + \beta|->|M_->|E_-> \equiv |\psi> \ . \qquad (6.3.3)$$

When the states of environment $|E_i>$ corresponding to states $M_\pm$ of the measuring device are orthogonal $<E_+|E_-> = 0$, the density matrix that describes the object + measuring device system obtained by tracing over degrees of freedom of the environment becomes diagonal

$$\rho_{OD} = \mathrm{Tr}_E \mid \psi >< \psi \mid = \mid \alpha \mid^2 \mid + > \mid M_+ >< + \mid < M_+ \mid + \mid \beta \mid^2 \mid - > \mid M_- >< - \mid < M_- \mid , \quad (6.3.4)$$

or

$$\rho_{OD} = \begin{pmatrix} \mid \alpha \mid^2 & 0 \\ 0 & \mid \beta \mid^2 \end{pmatrix} . \quad (6.3.5)$$

We may compare this expression with Eq. (1.6.26). And, as has been mentioned there, this density matrix does not present the results of the measurement, it only predicts possible results of the measurement. Though the density matrix is diagonal, it represents the improper mixture, i.e., it is not a mixture of states having definite values of the object and measuring device. It is the superposition (6.3.3) of these states. Eq. (6.3.5) is not equivalent to Eq. (6.3.3). Tracing over the environmental states, we neglect the correlations of type (1.6.25). The entropy of the state (6.3.3)

$$S = -\mathrm{Tr}\rho\ln\rho = 0 \ , \quad (6.3.6)$$

while the entropy of state (6.3.5) equals

$$S = -\left(\mid \alpha \mid^2 \ln \mid \alpha \mid^2 + \mid \beta \mid^2 \ln \mid \beta \mid^2 \right) > 0 \ . \quad (6.3.7)$$

Equation (6.3.6) follows from the fact that when the pure state density matrix is diagonalized, it has only one element (on the diagonal) which is equal to 1.

From the invariance of the trace under unitary transformations (and evolution of the density matrix in time can be regarded as a unitary transformation) it follows that entropy of the state of a closed dynamic system does not depend on the time

$$S = -\mathrm{Tr}\rho\ln\rho = \mathrm{const} \ . \quad (6.3.8)$$

In the general case, the density matrix ρ_{ODE} (of combined object + measuring device + environment system) will describe the results of a series of measurements (ensemble of measurements), provided it is diagonal (at least approximately), in all its indices

$$\rho_{ODE} = \begin{pmatrix} \rho_{0_1 D_1 E_1 ; 0_1 D_1 E_1} & 0 & 0 & 0 \\ 0 & \rho_{0_2 D_2 E_2 ; 0_2 D_2 E_2} & 0 & 0 \\ 0 & & \ddots & \vdots \\ \vdots & & & \\ 0 & 0 & 0 \cdots\cdots & \rho_{0_N D_N E_N ; 0_N D_N E_N} \end{pmatrix} . \quad (6.3.9)$$

It is obvious that this diagonal density matrix cannot be obtained from the initially pure state of the (6.3.3) type, by the time-dependent unitary transformation (6.2.6). It follows from the fact that the entropy of the pure state (6.3.3) equals zero, while the entropy of the state (6.3.9) has the order of the magnitude

$$S \approx -\sum^{N} \frac{1}{N} \ln \frac{1}{N} = \ln N \ , \tag{6.3.10}$$

provided 1/N is the order of the magnitude of the diagonal matrix elements of (6.3.9). It means that the transition from the pure state wave function of the combined system, object + measuring device + environment, to the density matrix of (6.3.9) type, cannot be performed by the unitary transformation, i.e., cannot be described by quantum mechanics (or it is an **R**-type process).

It should be mentioned that equation (6.3.9) is a sufficient condition of the measurement process. Logically, it does not preclude other options. Thus, due to the complexity of the environment, the initial (pure state) density matrix may evolve into an efficient (for all practical purposes) density matrix, describing the measurement process [183, 174]. Zeh [184] criticizes such an approach.

Another possibility is that the density matrix ρ_{ODE} of the ensemble of object (0) + measuring device (D) + environment (E) can be factorized (even approximately).

$$\rho_{ODE} = \rho_{OD}\rho_{E} \ , \tag{6.3.11}$$

while ρ_{OD} is the diagonal density matrix. This factorization (6.3.11) contradicts Eq. (6.3.8) (the entropy is the integral of motion)

$$S_{ODE} = S_{OD} + S_{E} = 0 \ . \tag{6.3.12}$$

It is clear that

$$S_{OD} > 0 \qquad \text{and} \qquad S_{E} \geq 0 \ . \tag{6.3.13}$$

The first inequality follows from the diagonality of ρ_{OD} (with the diagonal matrix elements other than 1 and zeros). Relation (6.3.13) contradicts the assumption that the whole system is described by the wave function, and its entropy is zero.

All the above considerations are based on the assumption that the environment is described by the wave function, i.e. it is in a pure state. Of course, such an assumption is quite arbitrary. More than that, the interaction of the macroscopic system with the environment causes the decoherence of the macroscopic system [175, 185, 49, 186, 187]. Decoherence destroys superpositions, the system looses its quantum character, and entropy of the system increases.

The environment can cause the decoherence, if the density matrix of the environment is itself diagonal, and corresponds to high entropy of the order of the magnitude of (6.3.10). As a matter of fact, in the above cited works on the decoherence it has been tacitly assumed that the environment has high entropy and is described by the diagonal density matrix. The same assumption about the density matrix of the dissipative system has been made by Fain [17, 18] (see section 2.3). (The case when the dissipative system is in the vacuum state requires a special consideration.)

Let us now consider a quantum system described by the wave function. The quantum system may be either the measured object or the object and measuring device together. This quantum system interacts with the dissipative system (environment). The latter system has a very large number of degrees of freedom and large entropy. In the presentation in which the density matrix of the dissipative system is diagonal, it has the form

$$\rho_{\alpha\beta} = \rho_{\alpha\alpha}\delta_{\alpha\beta} \ . \tag{6.3.14}$$

Having in mind the normalization condition

$$\sum_{n=1}^{N} \rho_{\alpha\alpha} = 1 \ , \tag{6.3.15}$$

and that all $\rho_{\alpha\alpha}$ have the same order of magnitude, we get

$$\rho_{\alpha\alpha} \sim \frac{1}{N} \ , \tag{6.3.16}$$

while the entropy has the order of magnitude (6.3.10), lnN, and

$$N >> 1 \ . \tag{6.3.17}$$

As a result of <u>irreversible</u> interaction with the dissipative system, diagonal elements of the dissipative system suffer negligible change of order (see Eq. (2.3.11))

$$\Delta\rho_1 \approx 0(V^2) \sim \frac{1}{N} \ , \tag{6.3.18}$$

and the off-diagonal elements (see Eq. (2.1.15)) obtain the order of the magnitude

$$\Delta\rho_2 \sim 0(V)\Delta\rho_1 \sim \frac{1}{N^{3/2}} \ , \tag{6.3.19}$$

and may be neglected ($N \to \infty$).

Diagonal and off-diagonal matrix elements of the density matrix of the quantum system (object) are initially ($t = 0$)

$$\rho_{nn}^{(0)} = |c_n|^2, \qquad \rho_{nm}^{(0)} = c_n c_m^* \ . \tag{6.3.20}$$

We consider the case when the relaxation of the off-diagonal matrix elements, resulted from the interaction with the dissipative system (relaxation time T_2), is much faster than the relaxation of the diagonal matrix elements (T_1)

$$T_2 << T_1 \ . \tag{6.3.21}$$

In this case, at time interval

$$T_1 >> t >> T_2 \ , \tag{6.3.22}$$

the wave function is collapsed. The resultant density matrix of the total system can be approximately presented as (6.3.9). The density matrix of the collapsed wave function at time (6.3.22) is

$$\rho_{nn}(t >> T_2) = |c_n|^2 \ ,$$

while the density matrix of the dissipative system suffer negligibly small change 1/N → 0. On the other hand, the change of the entropy of the quantum system is quite substantial (see, e.g. (6.3.7)); however, it is negligibly small in comparison with the entropy of the dissipative system

$$-\sum_{\ell=1}^{N} |c_\ell|^2 \ln |c_l|^2 << \ln N \ ,$$

where N is a number of degrees of freedom of the dissipative system (environment). (The entropy of the whole system does not change.)

The above scenario describes the collapse of the wave function, taking into account that the whole system cannot be described by the wave function, i.e. it is in a mixed state. Of course, this picture is much more realistic than that presented by the pure state of the whole system (including the environment). In the latter case the Schrödinger cat type paradoxes are inevitable. In the next sections the concrete examples will be presented.

In the case of small entropy of the dissipative system (e.g., vacuum state, spontaneous emission) the factorization approximation is not valid and the entropy is not additive. However, also in this case, the object (atom, molecule) suffers a substantial change, while each degree of freedom of the dissipative system, e.g. each mode of the electromagnetic field, suffers an infinitesimally small change.

6.4 Measurement of spins in the Stern-Gerlach experiment

As an example of the measurement process, we consider the measurement of the spin of atoms (or nuclei) by means of a Stern-Gerlach experiment, illustrated in Fig. 11. The motion in the z-direction occurring in the region of the inhomogeneous magnetic field can be neglected. It means that the velocity of atoms (in the x-direction) is high enough. The magnetic force gives the particle a momentum that is directed up or down, according to whether the spin is up or down ($s_z = \frac{1}{2}$ or $-\frac{1}{2}$). The resulting z-motion of the particle after it leaves the field carries it to a height that depends on the spin. In this way the observation of the position of the atom enables us to tell whether the spin is up or down.

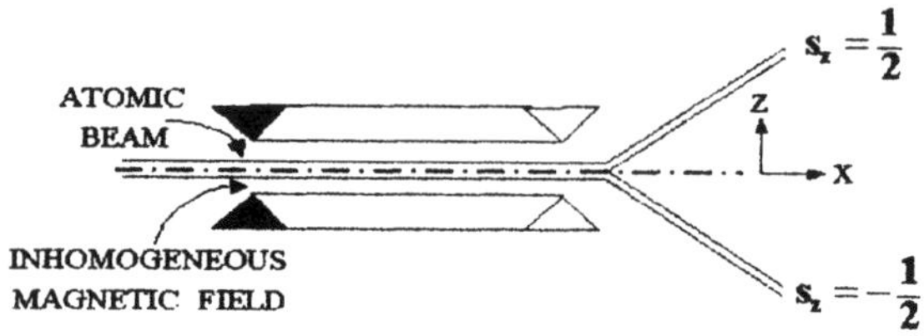

Fig. 11. Beam of atoms passes region of the inhomogeneous magnetic field. The resulting z motion of the atom after it leaves the field carries it to a height that depends on the spin.

The Hamiltonian of the interaction of spins of atoms with the inhomogeneous magnetic field may be presented in the form

$$H_I = -2\mu_0 s_z (H_0 + zH'_0) \quad . \tag{6.4.1}$$

Here

$$\mu_0 = -\frac{e\hbar}{2mc} \quad , \tag{6.4.2}$$

s_z is a projection of the spin on the z-axis, H_0 is a z-component of a magnetic field at point z = 0, and

$$H'_0 = \left(\frac{\partial H_z}{\partial z}\right)_{z=0} \quad . \tag{6.4.3}$$

The equation for the density matrix (in the region of the inhomogeneous magnetic field) has the form

$$i\hbar\frac{\partial \rho}{\partial t} = [H_I, \rho] \quad . \tag{6.4.4}$$

In this region of the inhomogeneous magnetic field, the kinetic energy $p_z^2/2M$ is neglected. We assume that the z-dependence of the wave-packet may be described classically, with its width satisfying

$$\Delta p \Delta z >> \hbar, \qquad \text{or} \qquad \Delta k \Delta z >> 1 \quad , \tag{6.4.5}$$

where $p = p_z$ is the momentum in the z-direction, while k is the corresponding wave vector.

The initial density matrix of the system (moving atom with the spin) may be written in the form

$$\rho_{II'}(z,z';0)=\sigma_{II'}P(z,z') \ . \qquad (6.4.6)$$

Here $P(z,z')$ - the density matrix of moving atoms (without taking into account their spins), while $\sigma_{II'}$ is the spin density matrix, I, I' being the eigenvalues of s_z $(\pm\frac{1}{2})$. The density matrix at the moment Δt of an exit from the region of inhomogeneous magnetic field, has the form

$$\rho_{II'}(z,z';\Delta t)=\sigma_{II'}P(z,z')\exp\{\frac{i}{\hbar}[2\mu_0 H_0(I-I')+2\mu_0 H_0'(zI-z'I')]\Delta t\} \ . \qquad (6.4.7)$$

To find the density matrix of the spin subsystem we have to trace density matrix (6.4.7) over z-variables

$$\sigma_{II'}(\Delta t)=\int dz\rho_{II'}(z,z;\Delta t)$$

$$=\sigma_{II'}\exp[\frac{i}{\hbar}2\mu_0 H_0(I-I')\ \Delta t]\int dzP(z,z)\exp[\frac{i}{\hbar}2\mu_0 H_0' z\ (I-I')\,\Delta t] \ . \qquad (6.4.8)$$

Let us consider the integral in the r.h.s. of this expression

$$\int P(z,z)e^{iKz}dz\,, \qquad K=\frac{2\mu_0 H_0'(I'-I)}{\hbar} \ . \qquad (6.4.9)$$

This is a Fourier component of $P(z,z)$. If K satisfies the condition

$$\frac{K}{\Delta k}=\frac{2\mu_0 H_0'(I-I')}{\hbar\Delta k}\Delta t \gg 1 \qquad (I\neq I') \ , \qquad (6.4.10)$$

the density matrix of the spin system becomes diagonal

$$\sigma_{II'}(\Delta t)=\sigma_{II}\delta_{II'}\int P(z,z)dz=\sigma_{II}\delta_{II'} \ . \qquad (6.4.11)$$

Similar to Eq. (6.3.4), the diagonal matrix (6.4.11) describes the improper mixture. Density matrix (6.4.11) does not describe an ensemble of already performed measurements. It only predicts results of the measurements. The spins do not have definite (but unknown values $\pm\frac{1}{2}$). There are correlations between the spin and moving atoms systems. These correlations are presented by Eq. (6.4.7). The density matrix, representing the proper mixture, has to be diagonal in all quantum numbers of the whole (the spin + moving atoms) system, as density matrix (6.3.9).

As a matter of fact, the density matrices (6.4.7) and (6.4.11) describe not yet separated beams ($I = \pm\frac{1}{2}$) of atoms. To consider the separation of the beams we will use the von Neumann equation

$$i\hbar\frac{\partial\rho}{\partial t} = \left[\frac{p_z^2}{2M},\rho\right] . \tag{6.4.12}$$

After atoms have passed through the magnetic field, the initial density matrix for Eq. (6.4.12) is

$$\rho_{II'}(z,z';0) = \rho_{II'}(z,z';\Delta t) . \tag{6.4.13}$$

The r.h.s. of this equation is determined by (6.4.7).

To solve Eq. (6.4.12) we will perform a transformation to the wave vector representation

$$\rho_{II'}(z,z',t) = \int dk \int dk' g_{II'}(k,k',t)\exp(ikz - ik'z') . \tag{6.4.14}$$

Substituting this integral into the von Neumann equation (6.4.12) we obtain

$$i\hbar\frac{\partial g_{II'}(k,k')}{\partial t} = g_{II'}(k,k')\left(\frac{\hbar^2k^2}{2M} - \frac{\hbar^2k'^2}{2M}\right) . \tag{6.4.15}$$

The solution of this equation has the form

$$g_{II'}(k,k';t) = g_{II'}(k,k';0)\exp\left[-\frac{i\hbar t}{2M}\left(k^2 - k'^2\right)\right] , \tag{6.4.16}$$

where $g_{II'}(k,k';0)$ is obtained from Eqs. (6.4.14) and (6.4.7)

$$g_{II'}(k,k';0) = \frac{1}{(2\pi)^2}\int dz \int dz' P(z,z')\sigma_{II'}$$

$$\times \exp\{\frac{i}{\hbar}[2\mu_0 H_0(I - I') + 2\mu_0 H_0'(zI - z'I')]\Delta t\}\exp(-ikz + ik'z')$$

$$= \sigma_{II'}P(k - k_I, k' - K_{I'})\exp[\frac{2\mu_0 i}{\hbar}H_0(I - I')] , \tag{6.4.17}$$

where $P(k,k')$ is the Fourier transform of $P(z,z')$ and

$$K_I = 2\mu_0\hbar^{-1}H_0' I\Delta t \tag{6.4.18}$$

Finally the solution of the von Neumann equation (6.4.12) satisfying initial condition (6.4.13) takes the form

$$P_{II'}(z,z';t) = \sigma_{II'} \int dk \int dk' P(k-K_I, k'-K_{I'}) \exp[\frac{2\mu_0 i}{\hbar} H_0 (I-I')\Delta t]$$

$$\text{x} \left[-\frac{i\hbar}{2M}\left(k^2 - k'^2\right) t \right] \exp(ikz - ik'z') \ . \qquad (6.4.19)$$

This density matrix describes two beams of atoms (with spins $\pm\frac{1}{2}$) after they pass through the magnetic field. Now, let us assume that spins (before the interaction with a magnetic field) were in the state with definite s_x - component, $s_x = \frac{1}{2}$

$$\psi'_{\frac{1}{2}} = \frac{1}{\sqrt{2}}\left(\psi_{-\frac{1}{2}} + \psi_{\frac{1}{2}}\right) . \qquad (6.4.20)$$

Here $\psi'_{\frac{1}{2}}$ is the eigenfunction of the spin-component s_x (with the eigenvalue $s_x = \frac{1}{2}$), while $\psi_{\pm\frac{1}{2}}$ are eigenfunctions of s_z. In this case the density matrix $\sigma_{II'}$ has the form

$$\sigma_{II'} = \frac{1}{2}; \quad \sigma = \frac{1}{2}\begin{pmatrix} 1 & 1 \\ 1 & 1 \end{pmatrix} . \qquad (6.4.21)$$

The off-diagonal elements of this matrix describe the interference between wave functions $\psi_{\pm\frac{1}{2}}$

$$|\psi'_{\frac{1}{2}}|^2 = \frac{1}{2}|\psi_{-\frac{1}{2}}|^2 + \frac{1}{2}|\psi_{\frac{1}{2}}|^2 + \frac{1}{2}(\psi^*_{-\frac{1}{2}}\psi_{\frac{1}{2}} + \psi_{-\frac{1}{2}}\psi^*_{\frac{1}{2}})$$

$$= \sigma_{-\frac{1}{2}-\frac{1}{2}}|\psi_{-\frac{1}{2}}|^2 + \sigma_{\frac{1}{2}\frac{1}{2}}|\psi_{\frac{1}{2}}|^2 + \sigma_{\frac{1}{2}-\frac{1}{2}}\psi^*_{-\frac{1}{2}}\psi_{\frac{1}{2}} + \sigma_{-\frac{1}{2}\frac{1}{2}}\psi_{-\frac{1}{2}}\psi^*_{\frac{1}{2}} \ . \qquad (6.4.22)$$

The density matrix (6.4.19) [with $\sigma_{II'}$, (6.4.21)] describe two correlated beams of atoms. These beams are similar to two beams of light in the two-slit experiment. Quantum correlations between two beams are described by off-diagonal elements of matrix $P(k,k')$.

Now there are three possibilities: (1) beams may be reunited, to interfere with one another; (2) a measurement will be performed on one of the beams, and (3) the beams will continue to depart.

The first possibility has been analyzed by Wigner [188]. Let us assume that two beams are reunited into a unique beam (see Fig. 12).

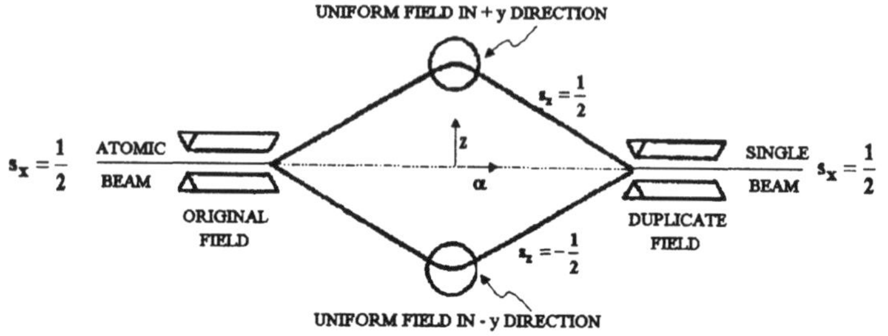

Fig. 12. The interference between the two beams of atoms in the Stern-Gerlach experiment is achieved by the arrangement of magnetic fields that reunite two beams together. Both beams have exactly symmetrical routes.

It is also assumed that some precisely devised combination of magnetic fields make two beams entirely symmetric. Then the interference of the beam in the region of the recombination revert the beams to the initial position. This means that the second Stern-Gerlach device oriented over the x-axis will measure, with the probability equal to one, $s_x = \frac{1}{2}$, i.e. the initial state (6.4.20). One can claim that measurement did not occur.

The second possibility may be realized by sending, let us say, one of the beams to the ion discharge camera. In this case the irreversible change has occurred, the beams cannot be reunited any more. The initial state cannot be reversed. In this case the Stern-Gerlach device, together with the ion discharge camera, perform the measurement. As a result of this measurement we find the spin in one of the states, $s_z = \pm\frac{1}{2}$. The observer does not interfere in this process. He may only verify the result of the measurement. One can easily realize that the irreversible change leads to the diagonal density matrix of the total system. The interference disappears irreversibly, as $t \to \infty$.

The third case (departing beams) is described by the density matrix (6.4.19) (see Fig. 13). Equation (6.4.19) describes both departure and spread of the beams. To find asymptotic and irreversible behavior of $P_{\mathrm{II'}}(z, z'; t)$, (6.4.19), we employ the formalism of section 1.8. We assume that the density matrix $P(k, k')$ has a sort of diagonal singularity [as in (1.8.5)]

$$P(k, k') = G(k, k')\delta(k - k') + h(k, k') \ , \tag{6.4.23}$$

where $h(k, k')$ has no δ-type singularities and is absolutely integrable. Then the asymptotic ($t \to \infty$) value of the density matrix (6.4.19) is [according to (1.8.6)]

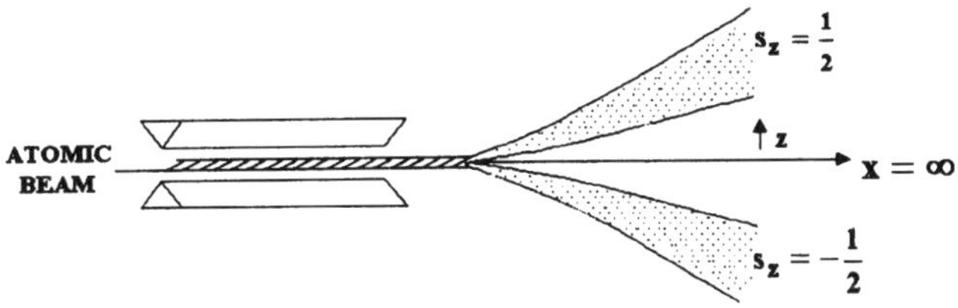

Fig. 13. Two departing wave packets with increasing spread and height as $t \to \infty$.

$$\rho_{\mathrm{II}'}(z,z';\infty) = \sigma_{\mathrm{II}}\delta_{\mathrm{II}'}\int dk G(k-K_{\mathrm{I}},k-K_{\mathrm{I}})\exp[ik(z-z')] \quad . \qquad (6.4.24)$$

Here we have neglected $G(k-K_{\mathrm{I}},k-K_{\mathrm{I}'})$ $(\mathrm{I} \neq \mathrm{I}')$, due to the assumption that two peaks of P(k,k) are far enough.

One should not worry that the density matrix (6.4.23) is not normalized, and its diagonal elements have singularities. The normalization can be performed by replacing the δ-function by a finite function that tends to the δ-function when a corresponding parameter tends to infinity. Then $P(k,k')$ is divided by $\sum_k P(k,k)$ and the parameter tends to infinity. The asymptotic density matrix in the k-representation has the form

$$P_{\mathrm{II}'}(k,k') = \sigma_{\mathrm{II}}\delta_{\mathrm{II}'}\delta_{kk'}G(k-K_{\mathrm{I}},k-K_{\mathrm{I}}) \quad . \qquad (6.4.25)$$

This density matrix describes the proper mixture of the (6.3.9)-type. It has to be emphasized that the density matrix (6.4.23) describes a mixed state which cannot be presented as the product of the functions of k

$$P(k,k') \neq c(k)c^*(k') \quad . \qquad (6.4.26)$$

In the density matrix (6.4.23) diagonal elements dominate. This circumstance conforms with general conclusions of the preceding section. The initially pure states may be decohered by the corresponding dissipative system (environment). Thus we come to the conclusion that asymptotically departing beams perform the measurement of spins $s_z = \pm\frac{1}{2}$. This measurement is described by the density matrix (6.4.25). The role of the observer is just to verify the accomplished measurement.

6.5 Realization of quantum measurement by the irreversible relaxation process

As has been shown in section 6.3, the interaction of the quantum system (having low or zero entropy), with a high entropy dissipative system, leads to the collapse of the wave function of the quantum system (which is initially in a pure state and has the wave function). As a matter of fact the relaxation of the dynamic system (section 2.7) may serve as a realization of quantum measurement [189].

In the factorization approximation (2.3.13), (2.3.14) and (2.3.17) the density matrix of the combined dynamic-dissipative system can be approximated by

$$\rho_{m\alpha;n\beta} = \sigma_{mn}(t)\rho_{\alpha\alpha}\delta_{\alpha\beta} \quad , \tag{6.5.1}$$

where $\rho_{\alpha\alpha}$ is the equilibrium density matrix of the dissipative system, while the equation for the density matrix σ_{mn} of the dynamic system satisfies equations (2.7.10) and (2.7.16)

$$\dot{\sigma}_{mn} + \frac{i}{\hbar}[E + \bar{V}^{ef} + \Gamma, \sigma]_{mn} = \begin{cases} -\sum_k \left(w_{nk}\sigma_{nn} - w_{kn}\sigma_{kk}\right), & m = n \\ -\left(\dfrac{1}{\tau_{mn}}\right)\sigma_{mn}, & m \neq n \end{cases} . \tag{6.5.2}$$

It is assumed here that the eigenfrequencies ω_{mn} of the dynamic subsystem are not degenerate. The rate coefficients w_{nk} and τ_{mn}^{-1} has the form

$$w_{nk} = \frac{2\pi}{\hbar}\sum_{\alpha\beta}\rho_{\alpha\alpha} \,|\, V_{n\alpha;k\beta} \,|^2\, \delta(F_\alpha - E_\beta - E_k + E_n) \quad , \tag{6.5.3}$$

(the transition probability per unit time) and

$$\tau_{mn}^{-1} = \frac{1}{2}\sum_k (w_{mk} + w_{nk}) + \frac{\pi}{\hbar}\sum_{\alpha\neq\beta} |\, V_{m\beta;m\alpha} - V_{n\beta;n\alpha} \,|^2 \quad , \tag{6.5.4}$$

and Γ is determined by Eq. (2.7.18).

A specific example of equations derived from Eqs. (6.5.2) are Bloch equations for the mean values of spin components s_x, s_y and s_z in the magnetic field $\vec{H} = \vec{k}H_0 + \vec{H}_1(t)$

$$\frac{d\vec{s}}{dt} = \gamma(\vec{s}x\vec{H}) - \mathbf{i}\frac{s_x}{T_2} - \mathbf{j}\frac{s_y}{T_2} - \mathbf{k}\frac{s_z - s_0}{T_1} \quad . \tag{6.5.5}$$

The so-called T_2 terms describe the phase relaxation, dephasing or decoherence, while T_1 terms describe the population relaxation. In Eq. (6.5.2), relaxation rates w_{kn} describe the population relaxation - "T_1^{-1}" terms, while τ_{mn}^{-1} describe the dephasing rates, "T_2^{-1}" terms. The pure dephasing is described by the second term of the r.h.s. of Eq. (6.5.4). The first term describes dephasing connected with the population relaxation.

We will consider the case of the fast dephasing

$$T_2 << T_1 \quad , \tag{6.5.6}$$

which is possible if the pure dephasing is much faster than the population dephasing. In the general case the phenomenon of fast dephasing should not necessarily be described by Bloch equations (6.5.5) or Eqs. (6.5.2). More general non-Markovian equations (2.4.1) may also describe relaxation accompanied by fast dephasing, symbolically represented by condition (6.5.6).

Asymptotically, the density matrix σ of the dynamic system becomes diagonal. As a matter of fact, the transformation of the initially off-diagonal density matrix of the dynamic system (as a result of interaction with a high entropy dissipative system) into a diagonal one is the characteristic feature of both the relaxation and measurement processes.

Now let us consider a dynamic subsystem E as a measured object which is characterized by the density matrix σ_{mn} ($c_n c_m^*$ in the pure case) at initial time $t = 0$, and let the dynamic system E be exposed to the relaxation process (interaction with the dissipative system) during the time interval Δt, satisfying the inequalities

$$T_2 << \Delta t << T_1 \ , \tag{6.5.7}$$

which are compatible with condition (6.5.6). Then, using Eqs. (6.5.1) and (6.5.2), one obtains at $t = \Delta t$

$$\rho_{m\alpha;n\beta}(\Delta t) = \sigma_{mn}\rho_{\alpha\alpha}\delta_{mn}\delta_{\alpha\beta} \ . \tag{6.5.8}$$

Thus during time interval Δt, the initial pure state (or general σ_{mn} – state) is transformed into the proper mixture (6.5.8). Since $\Delta t << T_1$, the diagonal density matrix elements σ_{nn} are unchanged, while for $\Delta t >> T_2$ the off-diagonal matrix elements vanish. The measurement process is distinct from the general irreversible process ($t \to \infty$), by the values of diagonal matrix elements. A general irreversible relaxation process also leads to the diagonal density matrix and proper mixture, but with $\sigma_{nn} = \sigma_{nn}^{(eq)}$, i.e. with both σ_{nn} and $\rho_{\alpha\alpha}$ being the equilibrium values of the dynamic E and dissipative systems density matrices. Therefore, relation (6.5.7) is a condition of the measurement: unchanged σ_{nn}, vanishing $\sigma_{nn}(1-\delta_{mn})$ and diagonal dissipative system density matrix $\rho_{\alpha\alpha}$.

6.6 Gedanken experiment: measurement of the z-component of spin ½

We will present here an example of a measurement performed by the relaxation process [189]. One can distinguish three stages in the measurement process: a preparation of the system (of course, it may be already prepared before the start of the measurement), the irreversible process of the measurement leading to the proper mixture and, finally, verification of the measurement, "looking at the pointer". The latter stage does not affect the system, the measurement has already been performed. The observer – the human being, does not intervene in the measured system. In the Gedanken experiment proposed here [189], the preparation is performed by the strong electromagnetic pulse which transfer the equilibrium state of the spin subsystem characterized by the diagonal density matrix $\rho_{nn}^{(eq)}$ and corresponding mean value of s_z

$$\overline{s_z} = \frac{1}{2}(\rho^{eq}_{\frac{1}{2}\frac{1}{2}} - \rho^{eq}_{-\frac{1}{2}-\frac{1}{2}}) \ , \tag{6.6.1}$$

(the pure state is characterized by $\rho^{eq}_{nn} = 1$ or 0) into the coherent state characterized by the off-diagonal density matrix. The irreversible measurement creates the proper mixture of $s_z = \pm\frac{1}{2}$ states with weights $\rho_{\frac{1}{2}\frac{1}{2}}$ and $\rho_{-\frac{1}{2}-\frac{1}{2}}$. A second electromagnetic pulse (a weak one, it does not change the state of the system) verifies that the system is in the proper mixture of states $s_z = \frac{1}{2}$ and $s_z = -\frac{1}{2}$.

Let us consider an equilibrium system of spins in a constant magnetic field H_0 directed towards a positive direction of the z-axis. The strong electromagnetic pulse "preparing" the state and the weak verifying pulse are directed along the x-axis. Then the Hamiltonian of spin $\frac{1}{2}$ in the magnetic field (constant and varying) has the form

$$H = \hbar\omega_0 s_z - 2\mu_0 s_x H_x \ , \tag{6.6.2}$$

where

$$\hbar\omega_0 = 2\mu_0 H_0 > 0; \quad \mu_0 = -\frac{e\hbar}{2mc} > 0 \ , \tag{6.6.3}$$

and we use the representation in which s_z is diagonal.

The time development of a spin system in the external magnetic field

$$H_x = H^0_x \cos\omega_0 t \, , \ V = V_0 \cos\omega t, \quad V_0 = 2\mu_0 H^0_x \ , \tag{6.6.4}$$

has the form

$$\dot{s}_z = -\frac{i}{2\hbar}(V_0 s_+ - V_0 s_-)\cos\omega_0 t - \frac{1}{T_1}(s_z - s^0_z) \ , \tag{6.6.5}$$

$$\dot{s}_+ = i\omega_0 s_+ + \frac{2\mu_0 i H_x}{2} s_z - \frac{1}{T_2} s_+, \quad s_- = (s_+)^* \ , \tag{6.6.6}$$

where s_z and $s_\pm = s_x \pm i s_y$ are mean values of spin operators, and T_1 and T_2 are introduced above relaxation times.

Now let us consider a simple case (the pure state) when the initial value of the spin in the ensemble of N spins is

$$s_z(0) = -\frac{1}{2} \quad . \tag{6.6.7}$$

This means that initially at t = 0 all N spins of the system were in their ground states, $-\hbar\omega_0/2$. We assume that the electromagnetic pulse duration is much shorter than T_2 and T_1. This means that one can neglect the relaxation terms with T_1 and T_2 during the action of the pulse. Then the solution of Eqs. (6.6.5 and 6.6.6) has the form

$$s_z = s_z(0)\cos\Omega_0 t; \quad s_x = s_z(0)\sin\Omega_0 t\sin\omega_0 t \ ;$$

$$s_y = -s_z(0)\sin\Omega_0 t\cos\omega_0 t, \qquad \Omega_0 = |V_0|/\hbar \ . \tag{6.6.8}$$

Now let us assume that a $\frac{\pi}{2}$ - pulse ($\Omega_0\tau = \frac{\pi}{2}$, τ is the duration of the pulse) is imposed on the spin system. Such a pulse prepares a new state of the system [distinct from the initial state (6.6.7)]. In this new state the mean value

$$\overline{s}_z = 0 \ . \tag{6.6.9}$$

This means that the probabilities of states $s_z = \frac{1}{2}$ and $s_z = -\frac{1}{2}$ are equal. However, this is not a proper mixture of states $s_z = \pm\frac{1}{2}$. This is a superposition of these states. The mean values of s_x and s_y (for N spins) are

$$\overline{s}_x = -\frac{N}{2}\sin\omega_0 t \ ; \ \overline{s}_y = \frac{N}{2}\cos\omega_0 t \ . \tag{6.6.10}$$

This is the superradiant state (section 4.5). The absorption (or induced emission radiation) depends on the phase difference between some probe (weak) electromagnetic field $H_x = H_x^0\cos(\omega_0 t + \varphi)$ and $\overline{s}_y$, Eq. (6.6.10) (see also Eq. (4.3.12)). This absorption is determined by equation

$$\dot{W} = 2\mu_0\omega_0 H_x \overline{s}_y \ . \tag{6.6.11}$$

The intensity of radiation in this state is proportional to N^2 (superradiance). I would like to stress that this state has certain well-defined physical features: the absorption depending on phase relations and superradiance. This state is not only characterized by certain propensities [190] or probabilities, it has a definite ontological status. To check the properties of this status we have to take the system of N spins, while N is not necessarily a very big number.

We now turn to the stage of the irreversible process: relaxation-measurement. We wait time Δt (after the π/2 - pulse is already over) satisfying condition (6.5.7). As a result of interaction with the dissipative system during the interval Δt, the diagonal matrix elements of the density matrix (probabilities of states $s_z = \pm\frac{1}{2}$) are unchanged, $\Delta t << T_1$, while the coherence expressed by Eq. (6.6.10) (or by the off-diagonal density matrix elements $\sigma_{\frac{1}{2},-\frac{1}{2}}$), is destroyed, $\Delta t >> T_2$. The spin system is transformed into a proper mixture of states $|s_z = -\frac{1}{2}>$ and $|s_z = \frac{1}{2}>$.

Now the spin system has an entirely different ontological status. The intensity of spontaneous radiation is proportional to N/2 – the number of spins in the upper states. The absorption rate is equal to zero. The spins in the ground states and the spins in the upper states give exactly the same contribution to the absorption, but with the opposite signs.

To summarize, at the first stage the system is prepared in the state (6.6.10). This is a superposition of $|s_z = -\frac{1}{2}>$ and $|s_z = \frac{1}{2}>$ states. The measurement is realized by the contact with the dissipative system during the time interval (6.5.7). The verification of the measurement can be performed by the interaction with a very weak electromagnetic field that does not affect the state of the system. Both spontaneous emission of radiation and absorption of radiation do not change the state of the system, provided that the time of interaction with the weak electromagnetic field is small enough.

CONCLUDING REMARKS

At first sight there is an irreconcilable contradiction between the irreversible master equation (2.7.10)

$$\dot{P}_n = -\sum (w_{nk}P_n - w_{kn}P_k) \ , \tag{1}$$

and reversible von Neumann equation (1.1.4)

$$ih\frac{\partial\rho}{\partial t} = [H,\rho] \ . \tag{2}$$

However, in the case when the energy spectrum of the quantum system is continuous, as been shown in section 1.8, the behavior of the quantum system becomes irreversible. This irreversibility can be understood if we consider the transition from a finite, bounded system to an infinite, unbounded system. For the finite system the quantum recurrence theorem (the quantum analog of the Poincaré theorem) states that the quantum system gets arbitrarily close to the initial state at arbitrarily large times. In the limit of the unbounded system the Poincaré recurrence time tends to infinity, and quantum movement becomes irreversible.

For large enough volumes of a condensed medium (or large resonator), the frequencies of its modes form a quasicontinuous spectrum. For example, in periodic structures the wave vector has quantized values

$$k_x = 2\pi n_x / A\ ; \quad k_y = 2\pi n_y / B; \quad k_z = 2\pi n_z / C \ , \tag{3}$$

where A, B, and C are the dimensions of the volume and n_x, n_y, and n_z are integer numbers. The energies of the modes are functions of these wave vectors

$$E(\vec{k}) = (n_{\vec{k}} + \frac{1}{2})\hbar\omega(\vec{k}) \ , \tag{4}$$

and thus they form a quasicontinous spectrum, if the dimensions A, B, and C are large enough. When $A \to \infty$, $B \to \infty$, $C \to \infty$, the spectrum becomes continuous. In this case the theorem, proven in section 1.8, states that quantum movement becomes irreversible.

In the case of the continuous energy spectrum one can deduce the transition probabilities (per unit time), which do not depend on time (section 1.7). These transition probabilities may serve as the coefficients w_{nk} of the master equation.

The consistent derivation of master equation (1), or other forms of generalized master equations, for the system with the continuous spectrum, is performed in Chapter II. Particularly the generalized equations may be Markovian, i.e. their coefficients do not depend on time. The example of such an equation is Eq. (2.4.14)

$$\dot{\sigma} = -i\hbar[E + \bar{V}^{ef}, \sigma] - \Re\sigma \ . \tag{5}$$

Here $\Re$ is a time-independent supermatrix, and $\bar{V}^{ef}$ is the time-independent averaged effective interaction energy. This equation, without the term $\bar{V}^{ef}$ is, sometimes, called the Redfield equation. The density matrix σ is averaged (traced) over the thermal bath variables density matrix ρ of the whole system

$$\sigma = Tr_B \rho \ . \tag{6}$$

The important approximation utilized in the derivation of Eq. (5) is the factorization approximation (2.3.17)

$$\rho_{m\alpha;n\alpha}(t) = \sigma_{mn}(t) \cdot P_\alpha \ . \tag{7}$$

It means that approximately the overall density matrix is the product of time dependent density matrix $\sigma_{mn}(t)$ with time time-independent equilibrium density matrix P_α. This approximation holds if the relaxation of the dissipative system is much fater than that of the dynamic system (2.3.13)

$$T_{diss} << T_{dyn} \ . \tag{8}$$

The fast relaxation of the dissipative system destroys the correlations between the dynamic and dissipative systems, once they are created due to interaction between dynamic and dissipative systems.

One of the astonishing discoveries of the last decades is that there are solutions of Eq. (5) which are not positive definite. In other words, Eq. (5) may lead to the non-physical negative probabilities. There is a simple explanation of this inconsistency. At times close to the initial $t = 0$, equation (5) is not valid ($\Re$ depends on time (2.4.8)). When the initial conditions are taken at a time interval in which Eq. (5) is not correct, the usage of this equation in the whole time axes, $t > 0$, is inconsistent. The situation is similar to the usage to the semiclassical approximation for description of the movement in the potential well. In the vicinity of turning points the semiclassical approximation is not valid, and a consistent solution is achieved by the procedure of matching with exact solutions near the turning points. Without the matching procedure one may get nonquantized solutions. The latter are non-physical, contradicting and not approximating the exact solutions.

Chapters III and IV are devoted to the irreversible processes (relaxation) connected with interaction with phonons and photons respectively. These processes may be considered as spontaneous and induced emission of bosons – phonons and photons. The specificity of the interaction with photons is in the existence relaxation via collective, superradiant spontaneous emission of photons, sections 4.5-8.

Chapter V is devoted to the memory effects, general relations, exact solutions, and comparison of various approximations: Markovian, rotating wave (RWA) and weak coupling. Such a comparison has been achieved considering the exact solution in the system comprising harmonic oscillator interacting with harmonic (boson) bath.

The book is devoted to the irreversibilities in quantum mechanics. However, it is shown, in Chapter V, that dissipationless regimes may emerge in a certain range of parameters of a generally dissipative system.

All the above irreversible processes take place in certain particular cases, which are characterized by the continuous energy spectrum. On the other hand, there are systems having the discrete energy spectrum – two-level system, harmonic and anharmonic oscillators, atoms and molecules. All these systems do not exhibit irreversible behavior – they are reversible. Nevertheless, the irreversibility of quantum measurement is ubiquitous in quantum mechanics, and refers to all quantum systems. In a sense the irreversibility lies in the heart of quantum mechanics.

Another point which has to be mentioned is the question of the possibility of quantum measurement. There are many proofs that quantum measurement is impossible. As a matter of fact it is claimed that quantum measurement is a process which cannot be described in the framework of quantum mechanics. Usually the proofs of impossibility of quantum measurement are based on the assumption that quantum object, measuring device, and environment are described by the wave function. Then, the entropy of the combined quantum system – the object, the measuring device and the environment, is zero and is conserved (does not depend on time). Therefore the reduction of the wave packet, which is accompanied by the increase of the entropy, contradicts the overall conservation of the entropy. On the other hand, the interaction of quantum object, described by the wave function, with another system – the environment, which is in a mixed state and described by the density matrix with high entropy, may lead to the reduction of the wave packet without essential relative change of the density matrix and entropy of the environment.

REFERENCES

1. Fain, V.M. (B.): *Photons and Nonlinear Media*, Sov. Radio, Moscow (in Russian), 1972.
2. Brilloin, L.: *Science and Information Theory*, Academic Press, New York, 1956.
3. Landau, L.D. and Lifshitz, E.M.: *Quantum Mechanics*, Academic Press, London, 1963.
4. D'Espagnat, B.: In *Preludes in Theoretical Physics in honor of V.F. Weisskopf*, eds. De-Shalit, A., Feshbach, M., and Van Hove, L, North-Holland, Amsterdam, 1966.
5. Golden, S. and Longuet-Higgins, H.C.: *J. Chem. Phys.* **33** (1960), 1479.
6. Fain, V.M. (B.): *Soviet Physics Uspekhi* **6** (1963), 294.
7. Bocchieri, P. and Loinger, A.: *Phys. Rev.* **107** (1957), 337.
8. Shulman, L.S.: *Phys. Rev.* **A18** (1978), 2379.
9. Shulman, L.S.: *Time's Arrows and Quantum Measurement*, Cambridge University Press, 1997.
10. Zwanzig, R.: *Lectures in Theoretical Physics (Boulder)*, Interscience Publishers, N.Y., p. 106, 1960.
11. Zwanzig, R.: *Physica* **30** (1964), 1109.
12. Nakajima, S.: *Progr. Theor. Phys.* **20** (1958), 1998.
13. Landau, L.D.: *Z. Physik* **45** (1927), 430.
14. Pauli, W.: *Festschrift zum 60. Gebürtstage A. Sommerfelds*, Hirzel, Leipzig, p. 30, 1928.
15. Bloch, F.: *Physik Z.* **29** (1928), 58.
16. Van Hove, L.: *Physica* **21**, (1957), 517; **23** (1957), 441.
17. Fain, B.: *Physica* **A101** (1980), 67
18. Fain, B.: *Theory of Rate Processes in Condensed Media*, Springer- Verlag, Berlin, New York, 1980.
19. Oppenheim, I., Shuler, K.E., and Weiss, G.: *Stochastic Processes in Chemical Physics*, The MIT Press, Cambridge, Massachusetts, Section 3.3, 1977.
20. Fain, B.M. (B.): *JETP* **42**, (1962), 1075; *Soviet Phys. – JETP* **15** (1962), 743.
21. Fain, V.M. (B.), Khanin, Ya.I.: *Quantum Electronics*, Pergamon Press, Oxford, Vol. I, 1969; Fain, V.M. (B.): *Photons and Nonlinear Media*, Sov. Radio, Moscow (in Russian), 1972.
22. Fain, B., Lin, S.H., and Wu, W.X.: *Phys. Rev.* **A40**, (1989), 824.
23. Romero-Rochin, V. and Oppenheim, I.: *Physica* **A135**, (1989), 52.
24. Romero-Rochin, V., Orsky, A., and Oppenheim, I.: *Physica* A156 (1989), 244.
25. Morillo, M., Cukier, R.I., and Tij, M.: *Physica* **A179** (1991), 41.
26. Wangsness, R.K. and Bloch, F.: *Phys. Rev.* **89** (1953), 728.
27. Bloch, F.: *Phys. Rev.* **102** (1956), 104.
28. Bloch, F.: *Phys. Rev.* **105** (1957), 1206.
29. Redfield, A.G.: *IBM J. Res. Develop.* **1** (1957), 19.
30. Fano, U.: *Phys. Rev.* **96** (1954), 869.
31. Tomita, K.: *Progr. Theoret. Phys. (Kyoto)* **19** (1958), 541.
32. Hubbard, R.: *Rev. Mod. Phys.* **33** (1961), 249.

33. Fain, V.M. (B.): *Sov. Phys. JETP* **9** (1955), 747; **6** (1958), 323.
34. Argyres, P.M. and Kelley, P.L.: *Phys. Rev.* **134** (1964), 98.
35. Chaturvedi, S. and Shibata, E.: *Z. Physik* **B35** (1979), 279.
36. Suarez, A., Silbey, R., and Oppenheim, I.: *J. Chem. Phys.* **97** (1990), 510.
37. Chang, T.-M. and Skinneri, J.L.: *Physica* **A193** (1993), 483.
38. Oppenheim, I., Shuler, K.E., and Weiss, G.: *Stochastic Processes in Chemical Physics*, The MIT Press, Cambridge, Mass., Section 3.3, 1977.
39. Kossakowski, A.: *Bull. Acad. Pol. Sci., Sêrie Math. Astr. Phys.* **20** (1971), 1021; **21** (1973), 649.
40. Gorini, V., Kossakowski, A., and Sudarsham, E.C.G.: *J. Math. Phys.* **17** (1976), 821.
41. Lindblad, G.: *Commun. Math. Phys.* **40** (1975), 147.
42. Alicki, R. and Lendi, K: *Quantum Dynamical Semigroups and Applications, Lecture Notes in Physics* **286**, Springer, Berlin, 1987.
43. Pechukas, P.: *Phys. Rev. Lett.* **73** (1994), 1060.
44. Pollard, W.T., and Friesner, R.A.: *J. Chem. Phys.* **100** (1994), 8359.
45. Kohen, D., Marston, C.C., and Tannor, D.J.: *J. Chem. Phys.* **107** (1997), 5236.
46. Louisell, W.H.: *Quantum Statistical Properties of Radiation*, Wiley, New York, 1973.
47. Haake, F.: *Statistical Treatment of Open Systems by Generalized Master Equations*, Springer-Verlag, Berlin, New York, 1973.
48. Haken, H.: *Laser Theory*, Springer-Verlag, New York, 1970.
49. Dekker, H.: *Classical and Quantum Mechanics of the Damped Harmonic Oscillator*, *Phys. Rep.* **80** (1981), 1-112.
50. Born, M. and Huang, Kun: *Dynamic Theory of Crystal Lattices*, Oxford, 1954.
51. Fain, V.M. (B.): *Izvestia VUZov, Radiofizika* **2** (1959), 167 (in Russian).
52. Lax, M.: *J. Chem. Phys.* **20** (1952), 1752.
53. Lin, S.H.: *J. Chem. Phys.* **44** (1966), 3759.
54. Bixon, M. and Jortner, J.: *J. Chem. Phys.* **48** (1968), 715.
55. Bixon, M. and Jortner, J.: *J. Chem. Phys.* **50** (1969), 3284.
56. Jortner, J. and Berry, R.S.: *J. Chem. Phys.* **48** (1968), 2757.
57. Englman, R. and Jortner, J.: *Mol. Phys.* **18** (1970), W2, 145.
58. Bixon, M. and Jortner, J.: *Adv. Chem. Phys.* **106** (1999), 35.
59. Markus, R.A.: *J. Chem. Phys.* **24** (1956), 966.
60. Levich, V.G.: *Adv. Electrochem. Electrochem. Eng.* **4** (1956), 249.
61. Förster, Th.: *Ann. Phys. (Leipzig)* **2** (1948), 55.
62. Dexter, D.L.: *J. Chem. Phys.* **21** (1953), 836.
63. Soules, T.F. and Duke, C.B.: *Phys. Rev.* **B3** (1971), 262.
64. Rackovsky, S. and Silbey, R.: *Mol. Phys.* **25** (1973), 61.
65. Abram, I.I. and Silbey, R.: *J. Chem. Phys.* **63** (19675), 2137.
66. Fain, B.: *Physica* **B+C, 106** (1981), 393.
67. Fain, B.: *Chem. Phys. Lett.* **67** (1979), 267.
68. Dekker, H.: *Phys. Rev.* **A44** (1991), 2314.
69. Dekker, H.: *Phys. Rev.* **E50** (1994), 4265.
70. Fain, B.: *Phys. Rev.* ***B26*** (1982), 5932.

71. Morse, P. and Feshbach, H.: *Methods of Theoretical Physics*, McGraw-Hill, New York, 1953.
72. Fain, B.: *Phys. Stat. Sol.* **(b) 63** (1974), 411.
73. Fain, B.: *Chem. Phys. Letts.* **56** (1978), 503.
74. Ginzburg, V.L. and Fain, V.M. (B.): *JETP* **39** (1960), 1323; *Sov. Phys. – JETP* **12** (1961), 923.
75. Lax, M.: *Phys. Rev.* **145** (1966), 110.
76. Fain, B.: *Phys. Rev.* **A24** (1981), 2685.
77. Van Vleck, H.: *Phys. Rev.* **59** (1941), 724.
78. Brya, W.J. and Wagner, P.E.: *Phys. Rev. Lett.* **14** (1965), 431; *Phys. Rev.* **157** (1967), 400.
79. Faughnan, B.W. and Strandberg, M.W.P.: *J. Phys. Chem. Solids* **19** (1961), 155.
80. Bron, W.E. and Grill, W.: *Phys. Rev. Lett.* **40** (1978), 1959.
81. Hu, P.: *Phys. Rev. Lett.* **44** (1980), 417.
82. Fain, B. and Lin, S.H.: *Physica* **128A** (1984), 164.
83. Fain, B. and Rojansky, D.: *Phys. Rev.* **B34**, (1986), 4323.
84. Pines-Rojansky, D. and Fain, B.: *Phys. Rev.* **B91** (1990), 2704.
85. Feynmann, R.P., Vernon, F.L. Jr., and Helwarth, R.W.: *J. App. Phys.* **28** (1957), 49.
86. Landau, L.D. and Lifshitz, E.M.: *Classical Theory of Fields*, Pergamon Press, Oxford, 1963.
87. Fain, V.M. (B.): *Sov. Phys. JETP* **6** (1958), 323.
88. Fain, V.M. (B.): *"Izv. VUZov", Radiofizika* **6** (1963), 207 (in Russian).
89. Weisskopf, V.: *Naturwissenschaften* **23** (1935), 631.
90. Ginzburg, V.L.: *Dokl. Akad. Nauk. SSSR* **24** (1939), 130.
91. Fain, B.: *Il Nuovo Cimento* **68B** (1982), 73.
92. Dalibard, J., Dupont-Roc, J., and Cohen-Tannouidji, C.: *Journ. De Physique* **45** (1984), 637.
93. Dicke, R.H.: *Phys. Rev.* **93** (1954), 99; Proceedings of the Third Quantum Electronics Conference, Paris, 1963, P. Grivet and N. Bloembergen (eds.), Columbia U.P., New York, 1969, p. 35.
94. Fain, V.M. (B.): *Uspekhi Fiz. Nauk.* **64** (1958), 273 (in Russian).
95. Andreev, A.V., Emel'yenov, V.I. and Hinskii, Yu. A.: *Sov. Phys. Usp (U.S.A.)* **24** (1983), 493.
96. Field, M.S. and MacGilivray, J.C.: *Coherent Nonlinear Optics, Recent Advances*, M.S. Field and V.S. Letokhov (eds.), Springer Verlag, Berlin, 1980, pp. 7-57.
97. Gross, M. and Haroche, S.: *Phys. Rep.* **93** (1982), 301.
98. Haroche, S. and Raimond, J.M.: *Adv. At. Mol. Phys.* **20** (1985), 347.
99. Leonardi, C., Persico, F., and Vetri, G.: *Riv. Nuovo Cimento* **9** (1986), 1.
100. Zheleznyakov, U.V., Kocharovskii, V.V. and Kocharovskii Vl.V.: *Sov. Phys. Usp.* **32** (1989), 835.
101. Fain, B. and Lin, S.H.: In *Multi-Photon Processes and Spectroskopy*, S.H. Lin, A.A. Villaeys, and Y. Fujimura (eds.), World Scientific, Singapore, London, 1995, p. 201.

102. Vrehen, Q.H.F. and Gibbs, H.M.: In *Dissipative Systems in Quantum Optics*, Bonifacio, R. (ed.), Springer Verlag, Berlin, 1982, p. 111.
103. Rehler, N.E. and Eberly, J.H.: *Phys. Rev.* **A3** (1971), 1735.
104. Bethe, H.A.: *Phys. Rev.* **72** (1947), 339.
105. Fain, V.M. (B.): *Sov. Phys. JETP* **9** (1959), 562.
106. Friedberg, F., Hartman, S.R., and Manassah, J.T.: *Phys. Lett.* **40A** (1972), 365.
107. Bloembergen, N. and Pound, R.V.: *Phys. Rev.* **95** (1954), 8.
108. Arecchi, F.T. and Courtens, E.: *Phys. Rev.* **A2** (1970), 1730.
109. Haake, F., King, H., Schroeder, G., Haus, J., and Glauber, R.J.: *Phys. Rev.* **A20** (1979), 2047.
110. Polder, D., Schuurmans, M.F. and Vrehen, Q.H.F.: *Phys. Rev.* **A19** (1979), 1192.
111. MacGillivray, J.C. amd Field, M.S.: *Phys. Rev.* **A14** (1976), 1169.
112. Scribanovitz, N.S., Herman, L.P., MacGillivray, J.C., and Field, M.S.: *Phys. Rev. Lett.* **30** (1973), 309.
113. Burnham, D.C. and Chiao, R.Y.: *Phys. Rev.* **188** (1969), 667.
114. Gibbs, H.M., Vreehen, Q.H., and Hikspoors, H.M.J.: *Phys. Rev. Lett.* **39** (1977), 547.
115. Vodopyanov, K.L., Kulevskii, L.A., and Voevodin, V.G.: *Optics Communications* **83** (1991), 322.
116. Belyanin, A.A., Kocharovsky, V.V., and Kocharovsky, VlV.: *Solid State Communications* **80** (1991), 322.
117. Vodopayanov, K.L., Kulevskii, L.A., and Gribenyukov, A.I.: *Journ. De Physique* **IV, Colloque**, Vol. 1 (1991), 391.
118. Belyanin, A.A., Kocharovsky, V.V., and Kocharovsky, VlV.: *Quant. Semiclass. Opt.* **91** (1997), 44.
119. Ginzburg, N.S. and Sergeev, A.S.: *Soviet Phys. – JETP Letters* **54** (1991), 446.
120. Belyanim, A.A., Kocharovsky, V.V., and Kocharovsky, Vl. V.: *Phys. Rev.* **E53** (1996), 5338.
121. Los' V.F., *Journ. of Physics, Condensed Matter* **3** (1991), 7027.
122. Ginzburg, N.S., Novoghilova, Yu.V., and Sergeev, A.S.: *Nuclear Instruments and Methods in Physics Research* **341** (1994), 230.
123. Andreev, A.V. and Polevoi, P.V.: *Sov. Physics – JETP Letters* **57** (1993), 107.
124. Andreev, A.V. and Polevoi, P.V.: *Quantum Electronics* **23** (1993), 863.
125. Knoester, J.: *Journ. of Luminescence* **53** (1992), 101.
126. Tokihiro, T., Manabe, Y., and Hanamura, E.: *Phys. Rev.* **B47** (1993), 2019; **48** (1993), 2773.
127. Fain, B.: *Quant. Semiclass. Opt.* **8** (1996); 429.
128. Fain, B.: *Physica* **A252** (1998), 461.
129. Goldberger, M.L. and Watson, K.M.: *Collision Theory*, Wiley, New York, 1964.
130. Rice, O.K.: *J. Chem. Phys.* **1** (1933), 375; Rice, O.K.: *Phys. Rev.* **33** (1929), 748.
131. Fain, B.: *Phys. Rev.* **A37** (1988), 546.

132. Fano, U.: *Phys. Rev.* **124** (1961), 1866.
133. Riess, I.: *J. Chem. Phys.* **52** (1970), 871.
134. Rosenfeld, J., Voigt, B., and Mead, C.A.: *J. Chem. Phys.* **53** (1970), 1960.
135. Gelbart, M. and Jortner, J.: *J. Chem. Phys.* **54** (1970), 2070.
136. Atabek, O. and Lefebvre, R.: *J. Chem. Phys.* **59** (1973), 4145.
137. Legget, A.J., Chakravarty, S., Dorsey, A.T., Fisher, M.P.A., Garg, A., and Zwerger, W.: *Rev. Mod. Phys.* **59** (1987), 1.
138. Harris, R.A. and Silbey, R.: *J. Chem. Phys.* **83** (1985), 5619.
139. Lee, T.D.: *Phys. Rev.* **95** (1954), 1329.
140. Davidson, R. and Kozak, J.J.: *J. Math. Phys.* **25** (1985), 556, and references therein.
141. Pfeifer, P.: *Phys. Rev.* **A26** (1982), 701.
142. Fain, B.: *Phys. Rev.* **B43** (1991), 8516.
143. Fain, B.: *Chem. Phys. Lett.* **172** (1990), 393.
144. Weisskopf, V. and Wigner, E.: *Zs. Phys.* **63** (1930), 54; **65** (1930), 18.
145. Agarwal, G.S., Haan, S.L., Barnett, K., and Cooper, J.: *Phys. Rev.* **A26** (1982), 2277.
146. Haus, J.W., Levenstein, M., and Rzazweski, K.: *Phys. Rev.* **A28** (1983), 2269, and references therein.
147. Coleman, P.E. and Knight, P.: *J. Phys.* **15** (1982), L235.
148. Greenland, P.T.: *J. Phys. B.: Mol. At. Mol. Phys.* **15** (1982), 3191.
149. Kofman, A.G., Kurizki, G., and Sherman, B.: *J. Modern Opt.* **41** (1994), 353.
150. Yablonovitch, E.: *Phys. Rev. Lett.* **58** (1993), 2059.
151. Yablonovitch, E., Gmitter, T.J., and Leung, K.M.: *Phys. Rev. Lett.* **67** (1991), 2295.
152. Ho, K.M., Chan, C.T., and Soukolis, C.M.: *Phys. Rev. Lett.* **65** (1990), 3152.
153. Kreuzer, H.J. and Gortel, Z.W.: *Physisorption Kinetics,* Springer, Berlin, 1986.
154. Gradshtein, I.S. and Ryzhik, I.M.: *Table of Integrals, Series and Products,* Academic Press, New York, 1994.
155. Fain, B.: *Phys. Rev. Lett.* **61** (1988), 2197.
156. Ondrechen, M.J., Nitzan, A., and Ratner, M.A.: *J. Chem. Phys.* **16** (1976), 49.
157. Lindenberg, K. and West, B.J.: *Phys. Rev.* **A30** (1983), 568.
158. Cukier, R.I. and Mazur, P.: *Physica* **53** (1971), 157.
159. Khidekel, V.: *Phys. Rev.* **E52** (1952), 2510.
160. Landau, L.D. and Lifshitz, E.M.: *Mechanics*, Pergamon, Oxford, 1976.
161. Khalfin, L.A.: *Soviet Physics JETP* **6** (1958), 1053.
162. Paley, R. and Wiener, N.: *Fourier Transform in the Complex Domain,* AMS, New York, 1934.
163. Bohr, N.: *Atomic Physics and Human Knowledge,* Science Edition, New York, 1961, pp. 63-64.
164. Albert Einstein, *Ideas and Opinions,* Deli Publishing Co., Inc., New York, 1979, p. 50.
165. Einstein, A., Podolsky, B., and Rosen, N.: *Phys. Rev.* **47** (1935), 777.
166. Einstein, A.: In *Albert Einstein, Philosopher Scientist*, P.A. Schilp, (ed.), p. 85, Library of Living Philosophers, Evanston, Illinois, 1949.
167. Einstein, A.: *Quantum-Mechanik und Wirklichkeit*, Dialectica II, 320, 1948.

168. Bohm, D.: *Quantum Theory,* Prentice-Hall, Inc., New York, 1952, Chapter 22, pp. 611-623.
169. Bell, J.S.: *Speakable and Unspeakable in Quantum Mechanics,* Cambridge University Press, 1993.
170. Clausner, J., Horn, M., Shimony, A., and Holt, R.: *Phys. Rev. Lett.* **26** (1969), 880.
171. Aspect, A., Granigier, P., and Roger, G.: *Phys. Rev. Lett.* **47** (1981), 460; *Phys. Rev. Lett.* **49** (1982), 91; Aspect, A., Dalibard, J., and Roger, G.: *Phys. Rev. Lett.* **44** (1982), 1804.
172. Weeler, J.A.: *Law Without Law*, in J.A. Wheeler and W.H. Zurek (eds.), *Quantum Theory and Measurement*, Princeton University Press, 1983, p. 182.
173. J.A. Wheeler and W.H. Zurek (eds.), *Quantum Theory and Measurement*, Princeton University Press, 1983.
174. Rosenfeld, L.: *Suppl. Progr. Theor. Phys.*, extra n. (1965), 222.
175. Zurek, W.H.: *Physics Today*, October 1991, p. 36.
176. S.F. Savitt (ed.), *Times Arrows Today*, Cambridge University Press, 1995.
177. Von Neumann, J.: *Mathematical Foundations of Quantum Mechanics*, Princeton University Press, Princeton, 1955.
178. Penrose, R.: *The Emperor's New Mind*, Vintage Books, London, 1989.
179. Schrödinger, E.: *Die Gegenwärtige Situation in der Quantenmechanik"*, Springer-Verlag, New York (1935) (English translation in Ref. 173).
180. Wigner, E.P.: *Symmetrics and Reflections*, Indian University Press, Bloomington, 1967; See, Ref. 173, Wigner, E.P.: *Remarks on the Mind-Body Question*, p. 168.
181. Peierls, R.E.: *Surprises in Theoretical Physics*, Princeton University Press, Princeton, 1979.
182. London, F. and Bauer, E.: *The Theory of Observation in Quantum Mechanics*, in Ref. 173.
183. Daneri, A., Loinger, A., and Prosperi, G.M.: *Nuclear Phys.* **33** (1962), 297.
184. Zeh, H.D.: *Foundations of Physics* **1** (1970), 69.
185. Feynman, R.P. and Vernon, F.L.: *Ann. Phys. N.Y.* **24** (1963), 118.
186. Caldeira, A.O. and Legget, A.J.: *Physica* **A121** (1983), 587; *Phys. Rev.* **A31** (1985), 1057.
187. Zurek, W.H.: In P. Meystre and M.O. Scully (eds.), *Frontiers of Nonequilibrium Statistical Physics*, Plenum, New York, 1985, p. 145; Unruh, W.G. and Zurek, W.H.: *Phys. Rev.* **D40** (1989), 1071.
188. Wigner, E.P.: *Am. J. Phys.* **31** (1963), 6.
189. Fain, B.: *Found. Physics Letters* **10** (1997), 221.
190. Popper, K.: *Quantum Theory and the Schism in Physics*, Routlege, London, 1995.
191. Jaynes, E.T. and Cummings, F.W.: *Proc. I.E.E.E.* **51** (1963)89.
192. Bohm, D. and Hilley, B.J.: *The Undivided Universe*, Routledge, London, 1993.

INDEX

FSC
www.fsc.org
MIX
Papier | Fördert
gute Waldnutzung
FSC® C083411

Kolibri 360 GmbH
Ferdinand-Jühlke-Straße 7
99095 Erfurt, Deutschland
produktsicherheit@kolibri360.de